Ghost Particle

Ghost Particle

In Search of the Elusive and Mysterious Neutrino

Alan Chodos and James Riordon

Foreword by Don Lincoln

The MIT Press

Cambridge, Massachusetts | London, England

The MIT Press would like to thank the anonymous peer reviewers who provided comments on drafts of this book. The generous work of academic experts is essential for establishing the authority and quality of our publications. We acknowledge with gratitude the contributions of these otherwise uncredited readers.

This book was set in ITC Stone Serif Std and ITC Stone Sans Std by New Best-set Typesetters Ltd. Printed and bound in the United States of America.

Library of Congress Cataloging-in-Publication Data

Names: Chodos, Alan, author. | Riordon, James R., author.
Title: Ghost particle : in search of the elusive and mysterious neutrino /
    Alan Chodos and James Riordon ; foreword by Don Lincoln
Description: Cambridge, Massachusetts : The MIT Press, [2023] | Includes
    bibliographical references and index.
Identifiers: LCCN 2022017646 (print) | LCCN 2022017647 (ebook) |
    ISBN 9780262047876 (hardcover) | ISBN 9780262373555 (epub) |
    ISBN 9780262373562 (pdf)
Subjects: LCSH: Neutrinos. | Neutrino astrophysics.
Classification: LCC QC793.5.N42 C45 2023 (print) | LCC QC793.5.N42 (ebook) |
    DDC 539.7/215—dc23/eng20220927
LC record available at https://lccn.loc.gov/2022017646
LC ebook record available at https://lccn.loc.gov/2022017647

10 9 8 7 6 5 4 3 2

To Louie and Abraham, in hope for the future.—A.C.

In memory of grandfather Clyde Lorrain Cowan Jr. With love for Mom, J.P., M.M., N.J., and little A.J.—J.R.

Contents

Foreword

Don Lincoln

I have something very important to tell you.

Now, I don't want you to be alarmed, but you are under a constant barrage by an insidious form of highly penetrating radiation. A powerful and uncontrolled nuclear reactor, generating an alarming 400 trillion trillion (4×10^{26}) watts of power, is spewing these subatomic particles in every direction and they are sleeting through you at the staggering rate of 500 trillion (5×10^{14}) every single second.

I'd tell you to seek shelter, but this form of exotic radiation is implacable; it can pass through the entire Earth with ease. You can run, but you cannot hide from this steady and relentless assault. After all, if the radiation can pass through the Earth, there is no way to shield yourself.

Frightened? That's understandable. What you've read sounds frightening. But you shouldn't be. The form of radiation I'm talking about is the benign neutrino, emitted when two protons in the Sun fuse together, ensuring among other things that your day at the beach will be pleasant. Despite the neutrinos' nuclear origin, they pose no danger at all. And you already kind of know that. If neutrinos pass through the Earth essentially never interacting, they certainly can pass through you. In fact, of the approximately trillion trillion (10^{24}) neutrinos that pass through you during a 70-year lifetime, perhaps a single one will interact in your body. It is for this reason that they are often called the ghosts of the subatomic world.

Because the neutrino interacts so little with ordinary matter, you should be forgiven if you asked yourself why you should be interested in it, but that would be a hasty question. In fact, the neutrino is one of the most fascinating of the denizens of the subatomic world. It has surprised researchers again and again, and it remains one of the least understood, with many outstanding questions.

Neutrinos were postulated in 1930, as a solution to a pressing mystery at the time. A form of radiation, called beta radiation, didn't seem to make sense. It seemed that energy was disappearing in the nuclear decay and that would have upended one of the cornerstone laws of physics—the law of conservation of energy. The neutrino was proposed to be light—perhaps massless—and to interact very weakly—weaker than any other subatomic particle.

The weakness with which neutrinos interact with matter means they are very difficult to detect; indeed, it took a quarter century and the taming of nuclear power for scientists to devise a clever enough method to produce and detect them. However, the validation of the existence of neutrinos was just the beginning. Just a year or two after neutrinos were observed, another team of researchers discovered that they were unique in the subatomic realm. While all denizens of the microcosm act as if they spin like a top, neutrinos and their antimatter counterparts were discovered to spin differently. It seems that neutrino interactions can distinguish between matter and antimatter.

And the surprises kept coming. A few years later, scientists found that there was not a single kind of neutrino but rather two, followed by the discovery of a third variant a quarter century later. The next puzzle was that measurements of the neutrinos emitted by the sun came up short, leading researchers to wonder if perhaps we didn't understand how the sun burned. Another option was that the three distinct types of neutrinos could morph into one another in a dizzying form of subatomic switcheroo, called *neutrino oscillation*. Neutrino oscillation was a crazy idea, until it was shown to be true near the start of the twenty-first century.

Even now, neutrinos have not revealed all of their secrets. While neutrino oscillation proved that neutrinos have a very small mass, scientists still don't know what the mass of a neutrino is. Indeed—and this is a true weirdness—the three types of neutrinos don't have a distinct and single mass, but rather a mix of three different values. This behavior is another property unique to neutrinos. Although researchers know something about the difference between the three different mass values, the absolute number still eludes us.

While scientists found that nuclear interactions involving neutrinos seemed to treat matter and antimatter neutrinos differently, the question of whether neutrinos and antimatter neutrinos are different or the same is still an open question. And researchers are building new facilities to study whether neutrinos and antimatter neutrinos oscillate in the same way or differently. If they differ, neutrinos may prove to be the reason why all of the matter of the universe exists. It's truly an exciting time to be studying neutrinos.

To be up to date in the field of particle physics means knowing about the past and future of neutrino research, and that's why *Ghost Particle* by Alan Chodos and James Riordon is such a timely and fascinating book. They delve deeply into the history of neutrino research, telling a series of gripping tales of scientific investigation. Chodos and Riordon tell of the letter that started it all. They regale us with plans to use an atomic bomb to search for neutrinos, followed by the less dramatic, but no less exciting, effort by researchers to use classified nuclear reactors to finish the job. They tell of Cold War defections and a missing-person case worthy of the best mystery writer. They divulge the details of an experiment searching for a handful of atoms of argon in an Olympic-size swimming pool of dry-cleaning fluid. And they tell of a future experiment, which involves firing a beam of neutrinos through the Earth to a detector about 1,300 kilometers (800 miles) away—a device built a mile underground in an abandoned gold mine and so huge that it will require the excavation of nearly a million tons of rock to build.

Ghost Particle doesn't limit itself to the successes of the past; it also muses about future possible uses of neutrinos, ranging from locating enemy nuclear submarines, hidden beneath the waves, to notifying national security officials of clandestine nuclear explosions. Even more speculative, they consider the idea of using neutrinos to search for intelligent extraterrestrial life. Admittedly, it's a long shot, but it's an approach that has not yet been investigated.

The neutrino, as exciting a particle as it has been in the past, has more tales to tell. Chodos and Riordon have written an enchanting and informative book, and the pages that follow are an excellent investment of time for anyone who wants to know more about this fascinating ghost of the subatomic world.

Preface

The world of our senses is only the surface of a deeper reality. The ancients conceived of elementary constituents like earth, air, fire, and water. Our modern roster of the basic components of matter began in the late nineteenth century with the discovery of the electron. The list has since proliferated. It includes three varieties (dubbed "flavors") of neutrino, perhaps the most mysterious of the elementary particles, and among the most difficult to study. As we shall describe, experiments on neutrinos, from their discovery in the 1950s to the present day and into the future, are heroic efforts demanding great skill, ingenuity, and perseverance.

Classic experiments have taught us much, and ongoing ones will teach us a lot more. It is a story about science, but, equally important, about the people who do science. Absent their intense dedication, none of the advances we describe would be possible.

Without using any mathematics more complicated than adding two numbers together, we also take a closer look at the physics behind the experiments. Neutrino physics involves some of the subtlest aspects of quantum mechanics. Indeed, the literature on neutrino oscillations, in which neutrinos change their flavor as they propagate from where they are produced to where they are detected, is filled with papers debating its finer points. We will give the reader a taste of the physics involved.

The realization, which crystallized about 20 years ago, that neutrinos have mass has prompted a series of urgent questions that physicists are eagerly trying to answer. Are neutrinos the same as their antiparticles? How big are their masses and how do they originate? Are neutrinos complicit in the generation of the matter in the universe, without which we would not exist?

These questions lead us naturally to explore the multifaceted role of neutrinos in astrophysics and cosmology. We also consider a variety of potential applications of neutrino physics to other areas of science and technology.

We do not cover everything that is known about neutrinos, for two reasons: First, there is so much happening at the present time that it can't all fit inside a single volume. We have therefore had to make some choices. And second, the field is progressing rapidly. Inevitably, some of what we describe will have been superseded even before the book reaches the bookstore shelves. But in the main, we have endeavored, as best we can, to address those topics that are of central importance to the neutrino story.

Portrait of a Ghost

Neutrinos, they are very small.
They have no charge and have no mass
And do not interact at all.

—John Updike, "Cosmic Gall"[1]

Neutrinos. They're tiny, just points in space. Each one barely affects anything else. They pass through you, the Earth, and the sun with hardly a trace. They almost never reveal themselves, but without them we couldn't know why the stars shine. Updike was wrong: Each has a minute mass. While their individual interactions are meager, Updike's "Cosmic Gall" overlooks their collective impact— there are so many that, all together, they might change the courses of galaxies. They are the most mysterious known particles in the universe. They are the ghost particles.

1
Neutrino Enigma

> This nothing-particle was named the neutrino and the only
> reason scientists suggested its existence was their need to make
> calculations come out even. . . . And yet the nothing-particle
> was not a nothing at all.
>
> —Isaac Asimov, *The Neutrino*[1]

Among the papers, photos, and graphs in a bin of materials that
neutrino pioneers Clyde Cowan and Fred Reines[2] collected, there
is a matchbox. It's still filled with matches that are now nearly 50
years old. On the top and bottom of the box are tidy notes inscribed
with a fountain pen. The message on one side of the matchbox
commemorates the 1972 solar neutrino conference held at the University of California, Irvine.[3] The other side proclaims that, in addition to as many as 30 matches, the box contains about 45 electron
neutrinos.

The commemoration suggests that neutrinos are flowing from
the sun in enormous numbers. At any given instant there are 45
solar neutrinos in every 25 cubic centimeter volume on Earth,
because that's the size of the matchbox, and the matchbox could be

Commemorative matchbox, on the occasion of the 1972 Solar Neutrino Conference, with handwritten notes indicating the number of solar neutrinos in a matchbox at any instant. The creator is unknown, but this was tucked among collected artifacts in the Cowan estate. *Source:* James Riordon.

Participants in Neutrino '72, the first formal conference dedicated to neutrino physics and astrophysics, held at Lake Balaton, Hungary, in June 1972. Sitting at left in the front row is T. D. Lee. Third from left is Richard Feynman, and Bruno Pontecorvo is fourth from left. Second from right is Clyde Cowan. Fred Reines is third from right, chatting with Victor Weisskopf, fourth from right. Ray Davis is standing second from upper left, next to John Bahcall in the striped shorts.

anywhere on the planet. Assuming you're about the size of a typical adult human, there are 100,000 or so neutrinos inside you right now that come from deep inside the sun.

Solar neutrinos aren't simply sitting in you. They travel at very nearly the speed of light. The ones in your body now are a snapshot of the 100 trillion solar neutrinos that pass through you every second.

The sun is one of the most intense sources of neutrinos that humans encounter, but not the only source by any means. Cosmic rays that stream down on our atmosphere produce showers of them. They flow out of nuclear reactors and emanate from naturally radioactive materials in the Earth. Commonly available wristwatches that rely on glowing tubes filled with tritium gas produce about a billion neutrinos per second.

There's one more supply of neutrinos that exceeds even the numbers of these—they come very nearly from the beginning of time.

We are immersed in neutrinos that are the oldest particles in existence. They're relics of the Big Bang and were created in the first trillionths of a second after the universe began. These relic neutrinos have cooled to just a few degrees above absolute zero and outnumber all other neutrinos. You have over three hundred Big Bang neutrinos in the tip of your pinky at this moment. When you take those into account, the matchbox holds far more than the 45 solar neutrinos noted on its label. Counting relic neutrinos, the number is closer to 8,000 in the 25 cubic centimeter volume of the matchbox.

There are more neutrinos in the universe than any other known particle except for the bits of light we call photons. And yet, since the neutrino was first hypothesized a little under a century ago, researchers have uncovered many more questions than they have found answers about the particle whose name means "little neutral one."

There are particles we are certain exist with properties that we know a lot about. Electrons, protons, neutrons, and photons are a few in a large collection of particles that scientists have identified, and characterized with startling precision, since the dawn of modern physics in the late 1800s.

On the other end of the spectrum, there are particles that might exist and that we know little about. Gravity-producing gravitons and various potential dark matter candidates are among as-yet-undiscovered particles that theorists have dreamed up to address various physics puzzles. If they exist at all, the hypothetical particles' properties are a matter of theoretical guesswork. Experiments so far have only managed to tell us what they're not. It's a little like homing in on Bigfoot by proving that he isn't in Times Square.

Neutrinos are distinguished by the fact that we're certain they exist, but, unlike other confirmed particles, we know comparatively little about them. The following are six established neutrino traits as of this writing:

- Neutrinos are fundamental particles with no internal parts.
- They have no electric charge.
- They have spin 1/2 (a quantum property that's difficult to visualize but is roughly like the spin of a top).
- They're subject to the weak force that's responsible for the radioactivity of some atoms.
- They come in at least three varieties, called flavors.
- At least some neutrinos almost certainly have mass.

By contrast, the following is a list of eight things we still don't know about neutrinos:

- How much mass do they have? Neutrinos have at most a millionth of the mass of an electron, which is the next lightest particle. All we know for certain is they are below the masses any experiment can measure so far.
- Which neutrino is heaviest? Neutrinos are blends of three masses. Two are close together, and the third is distinctly different. Whether the third is lighter or heavier than the other two

is still unknown. This is an important issue because it affects the predictions of theories and the outcomes of experiments that can fundamentally change our understanding of the universe.

- Where does their mass come from? The Higgs particle, discovered in 2012, confirmed the existence of a mechanism that endows other fundamental particles their mass. But we don't know why neutrinos have mass. It may be the Higgs mechanism, but that would imply that there are more neutrino types that no one has seen, which leads in part to the next mystery.

- How many types are there? There's no doubt that there are at least the three neutrino flavors. Theories that try to account for neutrino mass suggest there may be one or more types in addition to the ones we've already found, for a total of four, or perhaps six. Some experiments hint at one or more exotic neutrinos as well, potentially pushing the total number of neutrino types to many more than three.

- Are neutrinos their own antiparticles? Electrons, protons, and many other particles have distinct antimatter partners, with exactly the same mass but with opposite electric charge. Particle-antiparticle pairs can annihilate in a burst of energy, should they encounter each other. As particles without electric charge, neutrinos might be their own antiparticles, or maybe not.

- How are neutrinos connected to the existence of matter in the universe? If a neutrino is its own antimatter partner, it could help explain why there's enough matter to make galaxies, stars, planets. As a rule, anything that makes matter, like the Big Bang that started the universe rolling, also makes equal amounts of antimatter. But if matter and antimatter get together, they annihilate in a burst of energy. Neutrinos could hold the key to explaining why the universe has matter in it today, rather than being a vast, empty cosmos.

- Do neutrinos affect the dynamics of the stars and galaxies? Astronomers can tell from the orbits of stars at the outer edges of galaxies that the matter we can see is only about one-sixth of what must be there. The other five-sixths is mysterious *dark matter*. Neutrinos, both the ones we know and, especially, heavier ones that may await discovery, are among the possible candidates to explain dark matter.

- Do neutrinos have magnetic properties? We don't know how neutrinos respond to magnetic fields, if at all.

The picture is muddled still further due to conflicts between theory, experimental measurements, and astronomical observations. In a variety of cases, measurements and observations are, in turn, at odds with each other even as they contradict theory.

The situation is due to change. Experiments already underway, and others planned or under construction, will almost certainly settle many of the outstanding questions. Searches for some exotic neutrino types may come to fruition by the mid-2020s. It could take another decade or more to accurately measure neutrino masses. We may well have clear proof in a decade, if not sooner, indicating whether a neutrino is its own antiparticle. Discoveries such as these will shrink the list of remaining neutrino mysteries and fill out the catalogue of their characteristics, while potentially resolving some of the deepest puzzles in modern science. At the same time, forthcoming experiments could well unearth still more anomalies, contradictions, and conflict among different neutrino observations and experiments.

If the looming scientific discoveries are like many that have come before, the theoretical insights that will follow experimental outcomes will be concise, elegant, and powerful. In the view of Murray Gell-Mann, one of the architects of modern particle physics, such characteristics are among those that make theories beautiful. "We have this remarkable experience in this field of fundamental physics that beauty is a very successful criterion for choosing the right theory," Gell-Mann said in his 2007 TED talk.[4] "And when the mathematics is very simple—when in terms of some mathematical notation, you can write the theory in a very brief space, without a lot of complication—that's essentially what we mean by beauty or elegance."

Consider the periodic table of the elements. Prior to the discovery of subatomic particles, there was no coherent way to understand all the elements other than through a kind of chemical taxonomy. Early chemists were limited to sorting the elements the way a Victorian botanist categorized flowers, lumping them together based on their observable characteristics and chemical behavior. It led to a variety of organizational schemes, from simple lists in order of atomic mass that were rife with inexplicable gaps to almost mystical groupings relying on analogies to musical octaves. The arrangements were suggestive of something deep and important about elements.

Knowing the subatomic makeup of atoms allows the entire periodic table to fall into place. The concept of valences, which describes the way elements combine, was once based on observations alone. It can now be understood and predicted with our knowledge of the quantum nature of electrons. Isotopes of elements, with identical chemical properties but different masses, would be utterly inexplicable without knowledge of neutrons. And the nearly, but not quite, orderly increase of mass from element to element as you move along the periodic table can only be coherently explained once you know both about the particles in the nucleus and the forces that bind them together.

The anomalies and experimental tensions in neutrino physics may be a sign that we are in a position similar to that of chemists in the 1800s. There are hints from numerous other areas of science that our understanding of particle physics is incomplete. Pending neutrino discoveries alone are unlikely to rewrite physics the way subatomic particles rewrote the foundations of the periodic table. Neutrino-based mysteries do, however, offer tantalizing hints of potential cracks in some of the most successful and thoroughly tested theories the world has ever known. There's little doubt that the landscape of neutrino physics is on the verge of major shifts.

In the meantime, the same characteristics that make studying the neutrino challenging mean they offer a unique window into the Earth, stars, galaxies, and the universe itself. Neutrinos are born in nuclear reactions. Once they emerge, most glide through planets, stars, and the vast expanses of space while suffering no effect at all. Light, radio waves, X-rays, and other signals that astronomers use to study the cosmos are scattered, absorbed, and distorted by dust and objects between us and their sources.

It's difficult to catch neutrinos, but the ones that we manage to observe supply pristine information about the places of their origin—whether that's a nuclear reactor, the heart of the sun, a distant supernova, or even the very first moments after the Big Bang.

Neutrinos are already revealing things about the cosmos that seemed irrevocably hidden from us only a few years ago. They will soon help us to probe the universe in ways that are beyond other scientific instruments and techniques. Although practical applications of neutrinos still verge on fantasy, the neutrino's unique characteristics suggest that they could lead to commercial and technological applications that are simply impossible via any other means.

Before anyone could have imagined that it might have such potential, the neutrino became the central character in a decades-long drama. The first act would play out when it was still unclear whether or not neutrinos even existed.

2

A Desperate Proposal

Not everyone would be willing to say he believes in the
existence of the neutrino but it is safe to say that there is hardly
one of us who is not served by the neutrino hypothesis in
beta-decay.

—H. Richard Crane[1]

The neutrino was conceived out of conflict.

It was a mysterious little sibling to the electron, a new addition to
the family of particles that are the smallest, indivisible pieces that
everything in the universe is made of. And no one wanted it. The
very idea of neutrinos was a terrible thing, in the words of the person who first imagined it. Like a particle family curse, it was ultimately undeniable.

On the one hand was a puzzling discovery in the budding field of
atomic physics; on the other was a theoretical framework that had
emerged over the course of centuries. The solution to the clash was a
stark choice. It might be the ghostly neutrino that almost too perfectly
patched up the discord. Or we could cast aside laws that had guided
us from the mysticism of ancient philosophers to the atomic age.

For at least one of the leading physicists of the modern era, Niels Bohr, faith in the neutrino was too much to ask—the precious laws had to go.

Bohr's willingness to scrap the scientific cornerstones wasn't completely unreasonable. In science, after all, there are no sacred laws. Testing an existing law and finding that it's wrong, or discovering new limits indicating where it breaks down, are well-traveled routes to renown and respect in the sciences.

Some laws, however, are more resilient and fundamental than others. Conservation of energy and momentum are among the laws that have stood up to countless challenges and served as bedrock ideas for solving problems and testing theories. If physics ever had sacred tablets, they would be inscribed with laws covering various types of conservation.

Conservation laws describe things that are constant in a particular system. Energy, for example, can change form, but the total amount won't increase or decrease unless there's a connection to some outside energy source or a leak to let a portion of the energy escape.

A ball balanced on a hilltop has stored potential energy. As it begins to roll down the hill, some of that potential is converted into motion, or kinetic energy. At every moment as the ball careens down the hill, the sum of the potential and kinetic energies stays the same.

Similarly for momentum: It can be transferred from one object to another, as happens when billiard balls collide, but can't be created or destroyed. If you measure all the momentum on a pool table just before and after the break, conservation of momentum ensures the total will be constant.

In everyday life, energy and momentum conservation may not always appear to hold. That's only because it can be difficult to account for all the energy and momentum in a system. Billiard balls transfer some energy and momentum to the table surface. The table, in turn, is connected through its legs to the floor; the floor is on the Earth; the Earth orbits the sun, thanks to the gravity that binds them; and the moon orbits the Earth. And, of course, a pool player is at least momentarily connected to the balls via their cue stick. A perfect description of a billiard game would require keeping all those parts in mind.

There's the added complication that we're used to seeing things run down, as happens due to friction between the surface of a billiard ball and the felt-covered table, as well as the drag that comes from a ball moving through the surrounding air. A complete accounting would include every piece, along with the effects of friction and drag. These sorts of issues are part of the reason Aristotle argued that the natural state of some objects is rest—a view completely at odds with energy and momentum conservation.

A clear understanding of the importance of conservation laws came about only with the ingenuity of scientists such as Galileo and Newton, along with the realization that heat resulting from friction and drag is just another form of energy.

Advances over the last few centuries have led to conservation laws becoming the foundations for theories of motion, gravity, electricity and magnetism, chemistry, biology, and just about every other modern science. In all those fields, conservation laws are crucial for describing the world, making predictions, and checking experimental outcomes. If your new theory violates a bedrock conservation law, then the theory is almost certainly wrong.

Unless it's not, and a scientific revolution is underway. Starting around 1914, it looked like there was at least the possibility that some of the most esteemed conservation laws of physics might be crumbling.

Conservation Revolution and Crisis

Albert Einstein elevated the energy conservation law to new, more powerful heights in 1905 by showing that matter and energy are essentially the same thing. The total energy of a system includes the energy contained in the mass of its constituents, as well as any potential and kinetic energy.

We know precisely how the mass and energy are related based on Einstein's iconic equation that appears on everything from textbooks to T-shirts and bumper stickers: $E = mc^2$.

The equation shows that energy (E on the left of the equals sign) is the same as mass (m on the right) provided it's multiplied by the speed of light squared (c^2). With his equation, Einstein showed that conservation of mass is just an aspect of the more general conservation of mass/energy. In most sciences, and particularly in physics, replacing multiple laws with a single one is a sign of progress. It leads to clearer, more compact, and more elegant ways to understand the universe. Einstein's brief equation is among the most important scientific simplifications in history.

Suppose, for example, you have an atom, sitting still in space. If it happens to be a radioactive atom, it can blow apart at any moment. These atomic explosions are called "decays" in physics jargon. Like a decaying apple, it's change that happens on its own. For an apple, energy is released as microbes break it down. For an atom, energy is released when a piece breaks off of the nucleus at the core of the atom, or when it emits light.

If an atom happens to break into two pieces, Einstein's insight can help to tell you everything there is to know about the parts you end up with. You know the mass of the initial atom. As a result of Einstein's equation, you also know its total energy. After it breaks apart, the two pieces move away from each other with velocities you can measure, and you can also measure the masses of the two pieces.

If you add all the energy after the split, including both the energy in the masses of the pieces and the energy of their motion, it will exactly equal the total energy in the atom you started with. In fact, you don't even need to measure the energy of both components to know that total energy and momentum don't change when a particle breaks into two pieces. All you really need to do is check to make sure that one of the pieces always has the same energy every time the decay happens.

Consider firing a handgun. At first, you have one piece—the loaded gun—and after, you have two—the smoking gun and a bullet in flight. The bullet leaves the chamber with the same energy each time you fire it. If you measure the energy and momentum of the bullet (something gun aficionados do by firing into a block on a pendulum and checking to see how high it swings) you automatically know the energy and momentum transferred to the gun. This is essentially what the physicist Ernest Rutherford did when he discovered that the pieces coming out of radium atoms are alpha particles.

An alpha particle is identical to a helium atom, except without the electrons that normally accompany the helium nucleus. That is, it's a bare nucleus built of tightly bound pairs of protons and neutrons. At the time, physicists weren't aware that neutrons existed, and knew of only two particles that they believed to be the fundamental building blocks of all matter: electrons and protons. The logical assumption at the time was that the alpha particle was a new, subatomic equivalent of a bullet fired from an atom.

Every time that Rutherford measured the energy of alpha particles coming from the decay of radium, he found that they moved at the same speed, which means they had the same kinetic energy in every case.[2]

The situation was murkier with another particle that Rutherford studied. It was appropriately, if somewhat unimaginatively, called the beta particle because it was the second of three distinct types of radiation that Rutherford identified, along with alpha and gamma radiation. Soon after they were first discovered, beta particles were revealed to be electrons. Radioactive decays that produce electrons are still called *beta decays* today.

The decay appears to result in only two particles: a daughter nucleus left over after a parent nucleus decays, along with a single beta particle. As with the alphas, energy and momentum conservation demanded that the beta particle fly away with precisely the same speed with each decay, every time.

Except that's not what the experiments showed. Instead, British physicist James Chadwick found that the beta particles emerged with a range of speeds, which means they came with a spectrum of energies. The energy and momentum didn't add up. It was missing in amounts that varied from measurement to measurement, in an apparent breakdown of conservation. Something unexpected was going on. Just what wasn't exactly clear.

There were three primary competing ideas to address the puzzling conflict between the conservation laws and beta decays. Rutherford proposed that the beta particle escapes with the full energy required by conservation laws, but then lost random amounts of energy prior to the measurements taking place. How the energy disappeared, where it went, and why there was no sign of it weren't clear to Rutherford, or to anyone else for that matter. Atomic physics trailblazer Lise Meitner was among the supporters of the idea that the energy was lost as the beta particle moved through strong electric fields near the nucleus.[3] Niels Bohr had another, much more radical idea.

Bohr's stature as the father of quantum mechanics made his views on what might explain the beta decay spectrum worth serious consideration. In 1913, inspired by Rutherford's discovery of the atomic nucleus, Bohr had invented a model of the atom with a heavy nucleus surrounded by orbiting electrons. It worked well for the hydrogen atom, with one proton and one electron, and explained why the electron occupied discrete, or quantized, orbits in the atom. It was among the very first forays into quantum physics. While Bohr's model of the atom was revolutionary, it was tame in comparison to his proposal to explain beta decay a decade and a half later.

Bohr argued that the spectrum of energies of the emerging electron meant the conservation of energy just didn't hold for subatomic particles. He was advocating that it was time to abandon the conservation laws that were some of the most successful ideas in the history of science.

At first glance, Bohr's proposal might not have seemed particularly far out for the newborn field of quantum mechanics. It is, after all, a discipline fraught with phenomena that have no intuitive analogs in our daily lives. As we now know, things like quantum tunneling that allows particles to suddenly appear on the other side of an impenetrable barrier, entanglement of particles that are physically far apart but somehow remain a single entity, and quantum teleportation are par for the course in quantum mechanics, and crucial for the existence of technology underlying transistors and lasers. Perhaps energy conservation was only a quaint idea that was the result of our ignorance of the quantum world, rather than our understanding of classical physics.

Bohr's inclination to toss out energy conservation to explain beta decay was different. It suggested not just that quantum effects can seem inscrutable but that our understanding of the familiar world around us is wrong. It also flew in the face of a guiding idea Bohr had developed to ensure that any weirdness that physicists proposed for the quantum world should smoothly transition to the normalcy of our experiences in the macroscopic world. It had worked flawlessly in every other case of quantum and classical worlds colliding. But Bohr didn't let it stand in the way of his rejection of neutrinos and conservation.

Correspondence Principle Breakdown

One of the remarkable early developments in quantum mechanics was the discovery that light comes in discrete pieces—that is, light is quantized. Einstein proposed in 1905 that light is made of particles we now call photons. Arthur Compton verified it with particle scattering experiments in the early 1920s.

For hundreds of years before that, scientists had considered light to move in waves, like ripples on water or sound in air. It's a good way to think of light in many circumstances. When waves pass through a small hole, they spread out in what's known as a diffraction pattern. It's a phenomenon that's easy to see as ocean waves run up against a sea wall with a small gap in it, or around the corner of a jetty. In high school lab classes, it's often demonstrated for light by shining a laser through a pinole. Instead of creating a beam on the other side of the hole, the light spreads out in a pattern comparable to the diffraction patterns that water waves produce.

A photon passing through a small hole can't spread out if it's a particle—it may be deflected, but it remains a single particle and cannot create a diffraction pattern on its own, in the way that a beam of light waves would. The two models of light—as waves, on the one hand, and particles, on the other—seem to be fundamentally incompatible. How can light consist of discrete pieces in some situations and smooth waves in others?

The guiding idea that made sense of the seeming disconnect between the grainy quantum world made of particles and photons and the smooth continuity of the world we experience is the correspondence principle. It was, notably, the brainchild of Bohr himself.

As you zoom out from a quantum system with a small number of particles, the picture includes more and more particles and increasingly larger objects and systems. In the transition, quantum effects are progressively smeared out, until you get to a point where classical rules take over entirely and quantum weirdness disappears.

If you were to send many photons through the hole, one at a time, toward a screen on the other side, there is no way to predict where any one of them will end up. If you were to mark the spot where each landed, you would initially find what appeared to be a random collection of dots. Keep it up long enough, though, and a distinctive pattern will emerge. It is the same pattern that would result if you aimed a beam of light waves through the hole. The quantum calculations that predict the emergence of the pattern are very different from the classical rules that, nevertheless, predict exactly the same pattern. The ideas that go into each approach are entirely distinct, with quantum rules describing lumpy collections of individual particles, and classical rules applying to a smooth and continuous world. The consistency between the quantum approach and the wave description of interference patterns is a classic example of Bohr's correspondence principle.

Bohr's view of beta decay, by contrast, broke with the theme that had become par for the course in quantum mechanics. It wasn't just that beta decays violated conservation of energy in the quantum world; it was that we could use the effect to violate the bedrock rule of energy conservation on the macroscopic scale as well.

The issue stemmed from the rationale that beta decay should run backward as well as forward. That is, if a nucleus in an unstable carbon atom can spontaneously emit an electron and convert to nitrogen, then a nitrogen atom should also be able to capture an incident electron and convert to carbon. If the electron coming out of a carbon atom experiencing beta decay can have a range of energies, then so can the atom going into a nitrogen atom during inverse beta decay when an electron is captured.

Suppose, Bohr's argument suggests, you fired electrons at a nitrogen atom. The atom would occasionally absorb an electron to become an unstable carbon atom, and subsequently beta decay again by releasing an electron. The result is one electron going in and another coming out.

If you fired only low-energy electrons at the nitrogen nucleus, the emitted electron would still come out with a range of energies. It would be possible to turn a source of cool electrons into a supply of hot electrons. The starting point is a low-energy electron and a nitrogen nucleus, and the ending point is an identical nitrogen nucleus and a higher-energy electron. It would then be a simple matter to extract the energy from the outgoing electron and send it back in to run the process over again and collect the extra energy each time. It was the blueprint for a free energy machine.

It's a reflection of the magnitude of the problem that the beta decay spectrum presented in the early 1900s that Bohr would entertain such an unorthodox idea. Free energy systems, and the perpetual motion machines often associated with them, are so thoroughly disproven and debunked that they have become iconic examples of pathological science.

In fact, the U.S. Patent Office has special rule when it comes to perpetual motion and free energy machines—unlike other patent applications, you must send them a working prototype before they will even consider reviewing it. No one has managed to get their patent application for such a device reviewed at the Patent Office since the rule was established.

It's tempting to assume Bohr was suggesting another case of odd quantum behavior that reconciles with our everyday reality by way of the correspondence principle, and that he believed beta-decay free energy machines were thought experiments applicable only to inaccessibly small systems. In a paper published in 1934 in the *Proceedings of the National Academy of Sciences*, however, physicist Richard Tolman of the California Institute of Technology relates a discussion with Bohr indicating that the quantum pioneer believed beta decay undermined conservation of energy generally, not merely in the quantum world.[4] "Nevertheless," wrote Tolman, "as pointed out to me by Professor Bohr in conversation the ejection of electrons having a definite wide range of energies from nuclei all of which are alike would also involve a statistical failure in energy conservation," meaning the effect would not be erased in real world, macroscopic systems. "Under these circumstances we could then obtain an *actual net increase in energy*[5] by allowing a decomposition to take place and then rebuilding the original nuclei using electrons having lower energies than the average of those that were spontaneously emitted."

The idea of perpetual motion and free energy machines is such an abomination in the scientific community that it's likely only a physicist of Bohr's standing could have advocated it without causing irreparable harm to their reputation and career. Given a choice between accepting the only other plausible solution to the beta decay problem, proposed by Austrian physicist Wolfgang Pauli, and abandoning a centerpiece of modern physics, Bohr chose the latter.

Pauli's Desperate Remedy

In late 1930, Wolfgang Pauli's personal life was in turmoil.[6] His mother had committed suicide three years before. His unhappy marriage, only a year old, had broken up. He was drinking too much. Rather than attend a conference on physics in Tübingen, Pauli chose to go to a ball in Zürich. However, he did not neglect the conferees in Tübingen entirely. On December 4, he sent them a letter, jocularly written but with a very serious intent. The following are some relevant excerpts:[7]

> Dear Radioactives, ladies and gentlemen,
>
> . . . I have come upon a desperate solution to save the interplay . . . of statistics and the energy principle, in light of . . . the continuous beta spectrum. To be precise, the possibility that electrically neutral particles, which I would like to call neutrons, could exist in the nucleus, which have a spin 1/2, follow the exclusion principle and distinguish themselves moreover from photons by not moving at the speed of light. The mass of the neutrons would have to . . . not be greater than 0.01 of a proton mass. The continuous beta spectrum would be intelligible, if one assumed that with beta decay, a neutron would be emitted with an electron, such that the sum of energy of the neutron and the electron is constant. . . .
>
> At the moment I don't trust myself to publish anything about these ideas and turn first to you trustingly with the question about the possibility of an experimental proof of such a neutron, if the

neutron were to have the ability to penetrate like a gamma ray or had a penetration rate 10 times greater.

I admit that my solution from the start appears less probable because, if they existed, one would already have seen neutrons a long time ago. But only he who dares, wins. The esteemed predecessor in my position, Mr. Debye, throws a light on the seriousness of the situation of the continuous beta spectrum, when he said to me recently in Brussels: "O, one should best not think about it, just as one shouldn't think about [the] new taxes." Hence, we should seriously discuss every path for a solution. My dear Radioactives, may you test and judge. Unfortunately, I cannot be in Tübingen in person. I am indisposed because of a ball taking place in the evening between the 6th and 7th of December. . . .

Your most humble servant
W. Pauli

Pauli proposed to call his new particle the *neutron* in his letter, but it didn't stick. The name would be applied to another particle in 1932 when James Chadwick discovered the subatomic particle now known as the neutron.[8] Like protons, Chadwick's neutrons are about 2,000 times more massive than electrons. Pauli guessed that the particle he envisioned would be comparable in mass to an electron, leading Enrico Fermi to introduce the term *neutrino*[9] (Italian for little neutral one) instead.

Pauli's proposed particle has just the right properties to explain the beta decay spectrum, while remaining completely undetectable with the experimental physics methods available at that time. Unlike the electron or proton, it would carry no electrical charge. Because of its mass, it should travel slower than the speed of light. To be consistent with quantum theory, it must carry the quantum mechanical equivalent of angular momentum known as spin. Specifically, it should have spin 1/2, which ensures that beta decay complies with the laws of conservation of angular momentum in quantum mechanics.

While insightful, Pauli's particle brings with it a troubling question: Where did the neutrino come from? Perhaps, he argued, the atomic nucleus was built of protons, electrons, and neutrinos. Scientists would soon find that there are indeed electrically neutral particles in the cores of atoms, but they are Chadwick's heavier neutrons, not the low-mass particles that solve the beta decay conundrum.

Lise Meitner's alternative suggestion that the beta particle radiates energy as it escapes the atom was a particularly promising one. It preserved the quantum aspects of beta decay without either upending conservation laws or requiring the invention of a seemingly contrived and potentially mythical particle. More importantly, it was potentially testable. Meitner did just that.

Meitner put beta sources inside lead-lined containers to ensure that every bit of energy from the decays would be captured. She could then measure the temperature, and deduce the energy inside, as the container warmed up. According to Meitner's proposal, beta particles were all initially emitted at the same energy, then lost some on the way out of the atoms. The entire amount of energy from the initial beta particle emission would have to be in the container somewhere, regardless of its form, and contribute to its heating. If, instead, beta particles were emitted with a range of energies, or some of the energy was carried off by an ephemeral neutrino, then the container would warm less because there would be less energy trapped inside for heating in the first place.

As it turned out, Meitner found that the container was too cool to support her explanation for the range of electron energies. Instead, it was clear that some of the electrons were starting out with less energy than others, and none of them were losing it along the way out of an atom. Meitner couldn't save energy conservation in beta decay. That left the culprit as either the neutrino or the breakdown of energy conservation.

While Bohr and a handful of influential physicists continued to argue against the neutrino and energy conservation, within a few years support for the neutrino solution grew. Few physicists were willing to jettison a conservation law that was so useful and thoroughly verified.

Attempts to explain how neutrinos could be fundamental constituents of atoms still had problems. Where were the neutrinos before the beta decays? If they were inside an atom, as Pauli guessed, what held them there until the decay happened?

One attractive possibility dodged the issue by suggesting that neutrinos were simply created at the time of the beta decay.[10] It was an almost magical solution that lacked a clear explanation. It was little more than a physics fairy tale, although an admittedly seductive one.

The situation became clearer with two subsequent developments. First, the discovery of the neutron in 1932 eliminated the need to somehow cram electrons inside the nucleus to balance out some of the positive charge from protons. Second, in 1934 Fermi postulated a more tractable theory of beta decay using the recently developed quantum field theory, which allowed for the creation and annihilation of particles.

Fermi's theory marks the introduction of a new force of nature called the weak force. One of its primary roles is to explain how a neutron converts to a proton, an electron, and an antineutrino (or the related process of a proton converting to a neutron, a positron, and a neutrino). Fermi's weak force theory was a precursor of one of the main ingredients of modern particle physics theory.

There are several key components of particle physics that Fermi could not have known about at the time. But his theory fit so well with what was then known that Pauli's hypothetical neutrino soon became the default explanation for the beta-decay spectrum.

Fermi's innovative and mathematically daring use of quantum field theory is arguably the first application of a whole new kind of particle physics. Its success finally convinced Bohr to abandon his heretical views on the collapse of energy conservation.

As time went on, physicists broadly agreed that the explanation for the continuous beta decay spectrum had to be the neutrino, whether we would ever actually see one or not.

Ghost Hunting

Within a few months of Fermi's publication, nuclear physicists Hans Bethe and Rudolf Peierls published a brief, disheartening calculation in the journal *Nature*. It was inspired by both Fermi's theory and the discovery of artificially induced radiation, which resulted in beta decay that spits out a positron instead of an electron.[11]

By considering the typical amount of time it took for an unstable atom to decay and give off both a positive beta particle and a neutrino, assuming neutrinos existed, Bethe and Peierls could apply Fermi's theory to estimate how likely it was that the reverse reaction would take place. That is, instead of an atom emitting particles and becoming a different element when a neutrino was released, they imagined a neutrino striking an atom to transform one of the subatomic particles in its nucleus. An experiment relying on the reaction would definitively prove the existence of neutrino as a real particle.

Unfortunately, the result of the calculation was bleak. Bethe and Peierls found that the chances of a neutrino interacting with an atom were so small that one of them could typically travel through a light year of lead, a quarter of the distance to the star Proxima Centauri, before it has a good chance of hitting anything at all. That's a million times the average distance from the sun to Pluto. In other words, their calculation showed that the chances a neutrino coming from beta decay would pass through the entire Earth rather than stopping somewhere along the way is about a trillion to one. "If, therefore, the neutrino has no interaction with other particles besides the processes of creation and annihilation mentioned," Bethe and Peierls wrote at the end of their 1934 paper in the journal *Nature*, "one can conclude that there is no practically possible way of observing the neutrino."[12]

Any experimental effort to find the neutrinos that so conveniently explained the beta decay spectrum was doomed to failure.

There simply weren't any sources on Earth that could produce enough neutrinos to turn up in any imaginable experiment. It was a predicament that Pauli was not proud of. "I have done a terrible thing," neutrino hunter Fred Reines recalled Pauli saying, "I have postulated a particle that cannot be detected."[13]

Still, some experimentalists were undeterred by the long odds. Maurice Nahmias, of the Victoria University of Manchester, took detectors into the Holborn Tube Station in London in 1934.[14] He was acting under the assumption that neutrinos had a significant magnetic moment, which is a measure of their tendency to line up with magnetic fields. By descending into the station 30 meters below the ground, he hoped to screen out other sources of radiation that would interfere with the measurement. Unfortunately for Nahmias and his attempt, the neutrino magnetic moment is far smaller than he hoped, if it has one at all, and there were no signs of the particle in the London underground.

In the late 1930s, University of Michigan physicist Horace Crane conducted a search for neutrinos in bag of table salt.[15] He knew that when an isotope of radioactive sulfur breaks down through beta decay and converts into chlorine, it must emit an antineutrino. His inspiration came from the realization the reaction runs the other way as well. That is, if a chlorine atom absorbs a neutrino, it will be converted into a radioactive sulfur atom. The sulfur atom will then break down again later, releasing a signature beta particle that can be easily detected (along with another elusive antineutrino, of course). Table salt is a compound of sodium and chlorine, so Crane put a small amount of the beta-emitting material mesothorium in a three-pound bag of salt, waited three months, and looked for the radioactive sulfur. He found none.

As Bethe and Peierls had calculated, the likelihood of catching neutrinos in experiments of the types that Nahmias and Crane performed was vanishingly small. Nahmias had overly optimistic hopes for the neutrino's magnetic tendencies. And Crane's salt bag was many trillions of times too tiny to have a chance of catching a neutrino coming from a modest radioactive source.

The experiments of Crane, Nahmias, and everyone else hunting neutrinos at the time seemed to confirm that neutrinos were uncatchable. They would never be more than invisible sprites made up to ensure that everything balanced out in beta decays.

While the calculations of Bethe and Peierls were correct, their implicit assumptions—particularly about the sources of neutrinos that would be soon available for experimentalists—were premature. Bethe would become a key figure in creating intense neutrino sources that would help prove that the little neutral one was more than just a terrible thing that sprang from Pauli's imagination.

3

Project Poltergeist

Poltergeist (from German Polter, "noise" or "racket"; Geist, "spirit"), in occultism, a disembodied spirit or supernatural force credited with certain malicious or disturbing phenomena. . . .

—*Encyclopedia Britannica*

Los Alamos is a small town of about 13,000 people. It's perched on a plateau at the foot of the Jemez Mountains in northern New Mexico. Ponderosa pine forests stand to the west, cool deserts to the east, and the Rio Grande winds its way down from Colorado before turning east toward the Gulf of Mexico.

In the early 1900s, it was home to the Los Alamos Ranch School, a private institution for privileged boys. Their curricula included liberal amounts of horseback riding, hiking, hunting, and camping. It offered an idyllic escape for the few students whose families could afford it (including a young Stirling Colgate, future physicist and heir to the Colgate family toothpaste fortune).

The school was accessible only by rugged dirt roads twisting through canyon bottoms and clinging to sheer cliff walls. The nearest major city was La Villa Real de la Santa Fe de San Francisco de Asís, more commonly known as Santa Fe, 35 miles to the south-east. At a thousand miles from the West Coast of the United States and over 1,500 miles from the East Coast, Los Alamos offered wide open vistas of the mountains and canyons under skies that are clear up to 300 days a year.

The outskirts of the surrounding plateaus provided long sight-lines for guard towers and outposts on the lookout for approaching forces or aircraft. They were features that most likely appealed to the military officers that physicist Robert Oppenheimer introduced to the area during World War II.

It was the perfect location for a top-secret mission intended to bring the war to an abrupt end. The Manhattan Project would result in a weapon tens of thousands of times more powerful than any imagined before: the atomic bomb. In addition to being a device of unprecedented destruction, the bomb would also turn out to be the most intense source of neutrinos to ever exist on Earth.

To See a Ghost

There are two factors that can make particles like neutrinos more detectable: high energy and large quantities. As energy increases, the likelihood that a given neutrino will interact with matter also goes up. Larger numbers of neutrinos offset the difficulty of detecting any given one. As Bethe calculated, a single neutrino of the type coming from beta decay can pass through trillions of kilometers of solid matter with only a small chance of interacting at all. If, instead, you could somehow find a source that produced neutrinos with trillions of times the energy of the ones from beta decay, or one that put out trillions of times the number that a reasonably sized lump of radioactive material typically emits, the chances that one of them will have a detectable interaction in a modest volume of material are quite good.

In the 1950s, there was no known source of ultra-high-energy neutrinos. But the atomic bombs that Los Alamos scientists, including Bethe, were building depended on reactions that happened to produce copious neutrinos in precisely timed explosions.

In the case of the plutonium at the core of atomic bombs, six antineutrinos result from the chain of decays that follow each atomic fission. When an atomic bomb goes off, there are lots of them released in a fraction of a second. The first atomic bomb detonation, which Manhattan Project director Robert Oppenheimer code-named Trinity, produced a burst of 20 trillion trillion (20,000, 000,000,000,000,000,000,000) antineutrinos. They were ideal neutrino sources, for anyone with the security clearances and equipment required to perform experiments near a nuclear weapon test.

Even a kilometer from ground zero, this would mean that a million trillion antineutrinos would pass through a cubic meter of material immediately after a fission bomb detonation. Based on the Bethe and Peierls calculation, there would be an excellent chance of seeing at least one antineutrino in a small detector located thousands of meters from a fission bomb the size of the Trinity test. This was a fact noted by two physicists at the Los Alamos laboratory in the early 1950s: Fred Reines and Clyde Cowan Jr.

Pursuit of the Poltergeist

Fred Reines joined the Manhattan Project in 1944 at the invitation of the brilliant young theoretician Richard Feynman. Much of his early work included fundamental calculations about nuclear explosions, including optimal detonation altitudes to maximize shockwave damage on the ground, and methods of observing distant nuclear explosions as a result of the waves they create in the atmosphere. According to Stirling Colgate (one of the last of the Los Alamos Ranch School students, who returned to the New Mexico mountains as physicist), a large body of Reines's nuclear weapons calculations remained classified as late as 1988.[1]

By 1951, Reines had grown weary of bomb research. He convinced the Theoretical Division leader at Los Alamos to give him time to ponder fundamental physics problems he could work on. After a few months of contemplation, Reines found himself fixated on the enormous neutrino flux he knew must result from each nuclear detonation. After a brief conversation with Enrico Fermi, who was spending the summer in Los Alamos, and a few tepid words of encouragement from the elder physicist, Reines settled on the idea. He would go in search of neutrinos. Neither he nor Fermi had any idea how to do it, practically speaking, other than estimating that an as-yet-unspecified detector weighing a ton or more would be necessary.

It was a chance encounter during a layover in the airport in Kansas City, on route to a physics conference, that Reines had a chat with another Los Alamos physicist who happened to have the experimental skills to help solve the problem of building a neutrino detector.

Clyde Cowan had made his way to the lab in 1949, after an overseas stint in the army and returning to the United States to earn a PhD in experimental physics at Washington University in St. Louis. He soon became group leader of the Nuclear Weapons Test Division in Los Alamos, where he focused on methods to monitor nuclear explosions by observing the radiation they produce.

The two young scientists were in a unique position: a brilliant theorist and talented experimentalist with access to the abundant research and staff expertise assembled for one of the most ambitious scientific/military efforts ever attempted. More importantly, they had the chance to design an experiment to run in close proximity to a nuclear detonation. It was an unusual opportunity at the lab where national security was an overwhelming focus to step away from weapons development, and instead to address one of the biggest and most challenging puzzles in science at the time. Can we directly detect a neutrino, or is it only a convenient, though brilliant, figment of Pauli's imagination? Cowan and Reines set out to find the answer with an effort they light-heartedly dubbed Project Poltergeist.

They chose to focus on an interaction that would produce a signal in response to a neutrino changing the identity of a particle in an atomic nucleus. According to the reaction they had in mind, if a neutrino were to strike a neutron in an atom, it could result in the neutron converting into a proton and releasing an electron.

Alternatively, an antineutrino striking a proton could produce a neutron and the electron antiparticle, the positron. It wasn't clear at the time whether neutrinos and antineutrinos were different particles, or whether a neutrino was its own antiparticle, as some physicists suspected. Cowan and Reines chose to focus on the latter, antineutrino-proton interaction. Their assumption was that the particles coming from the chain of reactions that follow the fission of plutonium atoms in a bomb would likely release the antimatter version, and that the two types were distinct. It would turn out later to be a fortuitous decision.

El Monstro

The physicists planned to assemble a container lined with light-detecting photomultiplier tubes and filled with a fluid, known as a scintillator, which produces flashes of light when excited by radiation. The scintillator can't respond to a neutrino or antineutrino directly, but the fluid is made of compounds that are rich in hydrogen, and the antineutrinos can interact with the proton in a hydrogen atom. On the very rare occasion that an antineutrino is absorbed in the scintillator, it converts a hydrogen atom's proton to a neutron and a positron. The positron very rapidly annihilates with a nearby electron, producing a pair of gamma rays. The gamma rays, in turn, excite the scintillator, causing it to emit lower-energy photons that register in the light detectors on the inner wall of the container.

Although the neutrino flux coming from a nuclear weapon test is substantial, even kilometers from the epicenter, the physicists proposed installing their detector close to ground zero to ensure a clear signal. Even if placed close to the detonation, the detector would have to be enormous. The largest scintillator experiments up to that time used a liter or less of fluid. Cowan and Reines planned for a detector filled with a thousand times more. They nicknamed the design *El Monstro*.

To protect El Monstro from the destructive shock wave that accompanies a detonation, they would dig a 150-foot-deep shaft near the bomb, suspend the detector halfway down, and fill in the space above with dirt. On detonation of the bomb, explosives would sever the cables and allow the detector to fall freely for about two seconds as the shock wave passed. The detector would gather data on the way down, until it landed on a bed of feathers and foam rubber. To ensure the assembly descended without bouncing on a cushion of air trapped below it, they would have to pump the air out of the shaft just prior to the test. They would not know the results of the experiment immediately. It would take a few days for the radiation from the 20-kiloton bomb test to die down before they could dig out the shaft and pull up the tank.

It was, to say the least, an ambitious engineering and scientific plan, involving delicate equipment and exquisite timing in a system only tens of meters from the center of the most destructive force ever released by humanity. It's a testament to their creativity and playfulness that Cowan and Reines inserted a private joke in

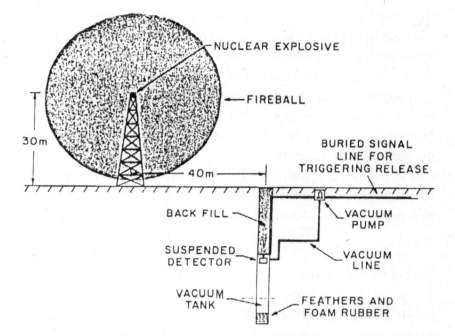

Cowan and Reines proposed suspending the El Monstro neutrino detector, filled with a ton and a half of scintillator fluid, in a shaft located 40 meters (about 137 feet) from an atomic bomb test tower. They planned to remotely release the detector when the bomb detonated and to collect data as it fell and the nuclear shock wave passed. *Source:* Los Alamos Scientific Laboratory.

the experimental layout. In most diagrams of the experiment, the distance from the tower that held the test nuclear weapon to the shaft that would contain the free-falling detector is shown to be 40 meters. The distance was arbitrary—anything up to a 100 meters or so would offer the neutrino flux the researchers needed. In fact, the distance they specified was approximately 137 feet, which is slightly over 40 meters.

The number 137 is intimately familiar to most physicists because 1/137 is the approximate value of an important number called the *fine structure constant*. It characterizes the electromagnetic forces between subatomic particles. Cowan would sometimes describe the choice of 137 feet as a good luck charm. At other times, he made jokes along the lines, "Where would one be if one walked from the tower to the experiment? One would be over about 137. . . feet away that is, which is a fine structure."[2]

The estimated El Monstro budget included $6,400 for the detector, $6,300 for the vessel to contain the detector, and $7,500 for drilling a hole in the Nevada desert to drop El Monstro into when the test bomb detonated. There was no line item in the estimate for the feathers and foam rubber for the experiment to land on. All told, it would cost just under $80,000 in today's dollars. It was a bargain price for an audacious experiment.

Killing the Monster

In late 1952, the drilling for the hole had already begun when Los Alamos lab physics division director J. M. B. Kellogg asked Reines to think it over one last time. Maybe, Kellogg suggested, they should reconsider conducting the experiment next to a nuclear reactor instead of a bomb.

Cowan and Reines had previously dismissed the approach. The high background radiation due to the perpetual rain of cosmic rays from above and the neutrons coming from a nearby reactor were the primary impediments. Both cosmic rays and neutrons lead to scintillator flashes that the detector wouldn't be able to distinguish from those emitted on annihilation of a neutrino-created positron with an electron. Cowan and Reines had opted for the extreme burst of neutrinos from a nuclear weapon to drown out all the unwanted noise for a fraction of a second.

Background interference would still be challenging, even with the enormous flash of neutrinos from a bomb. It was an issue Bethe had raised with Reines before. "I described," Reines would later write, "how, in addition to the use of bulk shielding which would

Omega West Reactor in Los Alamos Canyon, now decommissioned, decontaminated, and declassified. Tunnels dug 70 meters into the canyon wall behind the experimental fission reactor, and under the current town of Los Alamos, initially served as a high-security storage facility for the nation's nuclear weapons material stockpile. The rock above the tunnels provided shielding that allowed Cowan and Reines to test their neutrino detector components with reduced interference from cosmic rays. *Source:* Photograph by James Riordon.

screen out gamma rays and neutrons, we could use the delayed coincidence between the product positron and the neutron to identify the neutrino interaction."[3] That is, they could rely on the fact that every time an antineutrino interacted with a proton in their experiment, both a positron and a neutron would emerge.

Neutrons are a couple thousand times more massive than positrons, which means they lumber away more slowly than positrons after creation in inverse beta decay. While the positrons annihilate with nearby electrons in billionths of a second after they're created, the neutrons linger as they bounce off molecules in the detector fluid. After meandering about for a short time, which Cowan and

Reines estimated would be a matter of microseconds, the neutron is absorbed by the nucleus of an atom in the detector fluid and produces another gamma ray flash. An initial flash, followed by another about nine microseconds later, would be the signature of a neutrino reaction. Their reliance on the delayed coincidence measurements would allow them to discriminate between actual neutrino events and false signals that might come from other sources.

The technique presumably satisfied Bethe with regard to measuring neutrinos from a bomb, but the discussion with Kellogg inspired Cowan and Reines to revisit their calculations to see the impact of the dual flash approach on measurements near a reactor. "I have wondered since," Reines wrote, "why it took us so long to come to this obvious conclusion and how it escaped others."[4] The realization convinced them to dispense with the gargantuan El Monstro scintillator tank design and replace it with a 300-liter tank they called Herr Auge, a German phrase meaning Mr. Eye.

From Herr Auge to the Club Sandwich

In 1953, Cowan and Reines set up Herr Auge near a reactor that was part of the Hanford nuclear weapons production site in the south-central part of Washington State. It was a simple design, with 90 light-detecting tubes mounted through the walls of a cylinder filled with scintillating solution. The experiment initially showed a promising two and a half detections per hour. If the detector was working properly and counting neutrinos, the signal should have fallen toward zero when the reactor was turned off. Instead, they found that the detector signals hardly dropped at all when the reactor was not running. Despite the delayed coincidence measurement technique, the random background radiation was too high to clearly make out neutrinos in the data. The counts were statistically the same whether the reactor was running or not.

In 1953, Cowan and Reines share a laugh in the Project Poltergeist "Theoretical Division" office space adjacent to the heavily shielded Herr Auge detector, as it registered tantalizing hints of neutrinos coming from a nuclear reactor producing plutonium for nuclear weapons in Hanford, Washington. *Source:* Los Alamos Scientific Laboratory.

In 1953, they published a paper titled "Detection of the Free Neutrino," describing the experiment while acknowledging that the results, though tantalizing, didn't rise to the level of definitive detection.

Herr Auge had turned out to be a disappointment.

Still, Cowan and Reines found the tentative signs of neutrinos encouraging enough to return to Los Alamos and assemble a team to build a more sophisticated, better-shielded detector that might do the trick. The new design consisted of five layers, leading the team to call it the "club sandwich" in their formal paper in the journal *Science.*[5] Three of the layers, the "bread," were tanks filled with scintillator. Two intervening "meat" layers contained water with cadmium chloride dissolved in it. The essential antineutrino interactions would take place in the water (the meat) when a neutrino converted one of the protons in a water molecule to a neutron and a positron. The prompt annihilation of the positron would lead to two gamma rays moving in opposite directions, creating nearly simultaneous flashes of light in two scintillator layers (the bread) on either side. Microseconds later, when a cadmium atom captured the neutron, another burst of gamma rays would create flashes in two different slices of scintillator bread as well. The intensity of the light flashes indicated approximately the energy of the gamma rays, providing another check that allowed them to eliminate noise that could not have resulted from the annihilation or neutron capture, even if random flashes happened to occur with the correct timing.

The likelihood of random radiation producing signals with the coincidences and energy to mimic an antineutrino in the multi-layered Club Sandwich was much lower than in the comparatively simpler Herr Auge design, promising a significantly higher neutrino signal to noise ratio.

In 1956, the new detector was ready. The researchers loaded their Club Sandwich up and headed to a reactor in the U.S. Atomic Energy Commission's Savannah River Plant (now the Savannah River National Laboratory). The reactor facility outside Jackson, South Carolina, was producing plutonium for the nation's weapons program, and should have been putting out plenty of neutrinos as byproducts. But Cowan and Reines weren't alone in their pursuit of neutrinos at the plant.

The Race Is On

Shortly after the Club Sandwich was installed at Savannah River's P-reactor, Ray Davis of Brookhaven National Laboratory set up

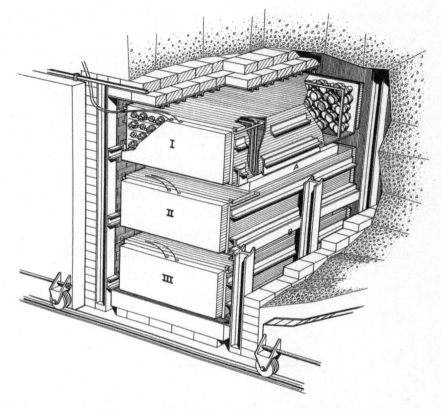

The Club Sandwich neutrino detector that Cowan and Reines installed near a reactor at the Savannah River nuclear site in 1956 consisted of three "bread" layers (labeled I, II, and III) filled with scintillator fluid, with collections of light-detecting photo-multiplier tubes at either end, and two "meat" layers (A and B) containing water with cadmium chloride dissolved in it. *Source:* Los Alamos Scientific Laboratory.

a detector at the facility's nearby R-reactor. Davis's experiment worked on fundamentally different principles from the Los Alamos collaborators' water and scintillator-based system. Instead, it relied on a reaction identified by prescient neutrino physicist Bruno Pontecorvo in 1946, who was then working at the Chalk River Laboratory in Canada.

In a report to the National Research Council of Canada's Division of Atomic Energy, Pontecorvo had explained how a neutrino could transform a chlorine atom into a radioactive form of argon.[6] If a large quantity of chlorine, or a chlorine-rich chemical, were placed close to a neutrino source, all a researcher would need to do is wait as the chemical was exposed to a neutrino flux, then examine the chemical for argon atoms. In practice, the experiment required a complex sequence of steps to pick out tens to hundreds of argon atoms, from a tank filled with three tons of chlorine-rich dry-cleaning fluid, after months of sitting next to a running reactor.

The Club Sandwich, on the other hand, provided evidence of neutrino interactions in real time, through traces registered on screens that could be photographed as the events occurred.

Their greatest advantage over Davis resulted not from experimental design, but instead from the guess that Cowan and Reines

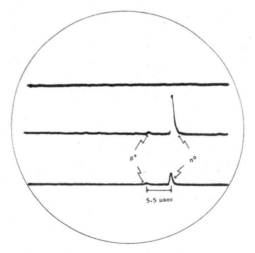

Oscilloscope traces show delayed coincidence signals indicative of neutrino-induced inverse beta decay. The traces are from light detectors in two of the Club Sandwich scintillator "bread" slices. The small, simultaneous pulses result from gamma rays produced when a positron annihilates with an electron. The gamma ray pulse coming 5.5 microseconds later, when a neutron is captured by a cadmium atom in one of the water-filled "meat" layers, is an indication that the positron came from inverse beta decay, rather than background radiation. *Source:* Los Alamos Scientific Laboratory.

had made years earlier—that neutrinos and their antimatter partners interact differently with matter, and specifically that the antineutrino variety were the ones produced in atom bombs and fission
reactors. The chlorine reaction Davis was using worked for the neutrinos, but not antineutrinos.

Sure enough, Cowan and Reines soon found that their experiment registered a clear signal of nearly three counts per hour. This
time, when the reactor was shut off, the signal fell significantly,
indicating the traces were real and not the product of false positives.

The team had suspected the reactions were due to antineutrinos,
but it wasn't until Davis searched for radioactive argon atoms in his
tank of dry-cleaning fluid, and found none, that the answer was
clear: Fission reactions produced antineutrinos, and they were different from the neutrinos Davis was after.

"On one occasion," said Davis at a 1979 memorial in honor of
the late Clyde Cowan, "they invited me over to their reactor, closed
the door, made me promise not to tell anyone, and then told me for
the first time that they were convinced that they had a positive signal of approximately the right magnitude. It was clear by then that
my experiment would not detect antineutrinos."[7] Still, Davis's null
result provided crucial information about the nature of neutrinos.
"My experiment showed that the neutrino was not its own antiparticle," Davis would later say in his 2002 Nobel Prize lecture.[8]

As we now know, neutrinos have tiny masses, and, as a result,
Davis's conclusion that the neutrino and antineutrino are necessarily distinct was premature. The neutrino may indeed be distinct from its antiparticle, as Davis assumed, or it may be that the
neutrino has a very small but nonzero probability of turning into
an antineutrino, and vice-versa, showing that they are in fact the
same. It's a question for experimentalists to answer, and is currently
the subject of active investigation.

Davis later repurposed his experiment to study the sun. It
would lead to a whole new missing-neutrino mystery, as well as its

ultimate solution. But Cowan and Reines had won the race to find the neutrino.

At long last, 42 years after Chadwick discovered the beta decay conundrum, 26 years after Pauli's desperately contrived solution, and 22 years after Bethe and Peierls had declared the particle entirely undetectable, the Cowan-Reines experiment proved conclusively that neutrinos are real. In 1995, the achievement was recognized with the only Nobel Prize ever awarded for research conducted under the auspices of the nuclear weapons laboratory at Los Alamos.

Enough Fun at Los Alamos

Cowan and Reines were eager to continue their investigations of the neutrino, which presented numerous additional puzzles, including some that remain unsolved today. The Los Alamos directors had other plans. In his memoirs, Reines would recall that when he and Cowan proposed follow-up experiments to their ground-breaking efforts, the reply they received was essentially, "You fellows have had enough fun. Why don't you go back to work?"

Cowan's response, after having tasted the thrill of neutrino research, was to resign almost instantly. He left New Mexico to take a position at George Washington University in Washington, DC. Reines persevered a little longer, but moved on two years later, when he went to the University of California at Irvine.

By the time the Nobel Prize for Project Poltergeist finally arrived, 40 years after the discovery, Reines was in ill health. He'd long abandoned any hope for the prize in recognition of their extraordinary experimental breakthrough. Cowan had died of a heart attack in 1974. Because Nobels are not awarded posthumously, the experimentalist of the team missed out on the honor entirely.

Nobel Prize nominations remain sealed for at least 50 years. Records now reveal that Cowan and Reines were nominated for a physics prize in 1957, the year immediately after they published their discovery of the neutrino, and every following year for which Nobel nominations have been unsealed.[9]

Delays between discoveries and Nobel Prizes are often lengthy, and the reasons are generally unclear, but with regard to Reines the case is particularly mysterious. Recognition in the scientific community of the significance of the initial neutrino discovery was nearly universal. The discovery of the muon neutrino, six years after the Project Poltergeist success, had already earned a Nobel for Leon Lederman, Jack Steinberger, and Melvin Schwartz in 1988.

The inherent secrecy of the nuclear weapons program, which led to an eight-year gap before the results of the initial electron neutrino discovery could be independently confirmed, was one possible roadblock on the way to a Nobel Prize. "I suspect," Reines wrote in his 1982 personal recollections of the neutrino hunt, "that the unseemly delay was due to the fact that our result was not unexpected but it may also have had to do with the initially highly classified nature of the neutrino source."

The experiment's roots in the nuclear weapons program, which had provided Cowan and Reines with a rare combination of resources to prevail in a search that had eluded so many for so long, may also have impeded the recognition. Alfred Nobel himself had grown rich from the development of explosives responsible for enormous carnage on the battlefield. After learning of the accidental publication of his premature obituary and finding that it ignored his hundreds of non-bellicose inventions, focusing instead on his contributions to military weaponry, Nobel experienced a personal epiphany. He bequeathed the bulk of his wealth to establish the world's most widely recognized scientific prizes in an attempt to redeem his legacy.

It is, perhaps, unsurprising that the Nobel committee was reluctant to bestow a prize on research so closely associated with the invention of nuclear weapons, and performed by researchers whose primary efforts focused, at least initially, on increasing weapon yields and optimizing their lethality.[10]

Today, the Herr Auge apparatus is the final physical remnant of the early efforts at Los Alamos to prove the neutrinos were real particles, rather than a mathematical sleight of hand. It still bristles with light detectors mounted in the wall of a cylindrical tube about the size of a modest dining table. It is locked away in the warehouse of the Lab-run Bradbury Science Museum in the center of the town outside the security gates. All available museum floor space at the Bradbury is instead dedicated to nuclear weapons and national security exhibits, with none left over for the most celebrated, purely scientific achievement in the history of Los Alamos National Laboratory.

As a final irony, a copy of the Nobel Prize medal bestowed on Reines in 1995 is now on a pedestal at the center of a cottage in a Los Alamos neighborhood that once was home to the lab's top scientists and administrators. Bathtub Row, as the area is still known, featured houses built with fully equipped bathing facilities, unlike most of the housing that lower-level employees occupied in the early days of the nuclear weapons lab.

The Hans Bethe House Museum is open to the public and honors one of the town's most famous residents, who happens to be the very person who so convincingly argued in 1934 that neutrinos from beta decays could never be detected. Although Bethe and Peierls were correct at the time that they did their calculation, given the neutrino sources and experimental techniques then available, Bethe was among the most prominent people whose work on the fission bomb resulted in the circumstances that made the Cowan-Reines experiment possible at all. Reines eventually had the chance to ask Bethe about his calculation asserting the neutrino's inherent undetectability. "Well," Bethe responded, "you shouldn't believe everything you read in the papers."[11]

4

Adrift in a Particle Sea

Plurality should not be posited without necessity.

—William of Ockham[1]

We are immersed in a sea of particles, infinite in number and extending throughout all the universe. For the most part, it is entirely invisible to us.

At least, that was the inescapable implication that confronted Paul Dirac when he developed a new theory in the early 1930s to describe the quantum physics of light and matter. To Italian physicist Ettore Majorana,[2] a gifted and emotionally fragile young protégé of Enrico Fermi, it was a bizarre complication to an otherwise brilliant theory.

Dirac's theory applied, in particular, to electrons moving freely in space. You can always add energy to an electron's motion, and make it move a bit faster. As you might imagine, there's no upper limit to the energy an electron can take on, but it requires some work to boost it up. An energetic electron will, from time to time, emit photons and lose some energy on its own to fall back down to a lower level.

A problem arises in Dirac's theory because, for every positive energy solution, there is a corresponding solution with negative energy.[3] Just as the positive energy solutions are unbounded above, and an electron could always be lifted to higher levels, the negative energy solutions are unbounded below, and an electron should be capable of falling into them. Dirac realized that a positive-energy electron could emit a photon and fall into a negative energy state. Because there is always a lower level for it to fall into, it would repeat the process over and over again, losing energy each time and producing more and more photons, in an endless cascade that releases an infinite amount of energy. It would be an energetic catastrophe that would be repeated for every free electron. Our existence in a relatively chilly universe with no signs of electron-induced cosmic conflagrations means these catastrophes aren't happening.

To resolve the discrepancy between his theory and reality, Dirac came up with the idea that all the negative energy states in the infinitely deep negative energy sea are filled. Once a state is filled, no other particle can occupy it. So, there is no longer any danger of a positive energy electron falling into the sea. There's simply nowhere for them to fall to. The problem is that to accept Dirac's otherwise brilliant theory, you must believe in the infinitely deep negative energy sea.

As a rule, we can't see any evidence of negative energy electrons. That is, the negative energy sea that surrounds us at all times, extends throughout all space, and contains an infinite number of electrons with energies extending to negative infinity is completely invisible and inaccessible to us almost all the time because it is completely filled up. According to the theory, we can become aware of the sea only when one of the negative energy electrons somehow acquires enough positive energy to pop out of it, leaving a hole behind. The newly liberated particle is like any other electron in the observable world, with positive energy and the normal amount of electrical charge. The hole that the electron leaves behind is its mirror image. It carries the opposite charge of an electron and moves about freely in the negative energy sea, except that, being a deficit of negative energy, it now also has positive energy, which makes it observable to us.

Dirac initially speculated that protons could be manifestations of holes left by electrons escaping the negative energy sea, but Enrico Fermi pointed out that the hole should have exactly the same mass as the electron that previously filled it, roughly two thousand times smaller than the mass of a proton. That is, a hole in the negative energy sea, according to the interpretation of quantum electrodynamics (QED) that Dirac developed, must be a new kind of particle. It had to be an antielectron.

Dirac's invisible sea of negative energy electrons with positively charged holes moving through it may seem bizarre, but the discovery of the electron's antimatter partner, which we now call the positron, a couple of years after Dirac initially developed his theory was a triumphant confirmation of the concept, regardless of whether the negative energy sea is real or a physics fairy tale. The positron had all the properties of a hole in the negative energy sea, even to the extent that a free electron encountering one would release a burst of energy with both particles disappearing, apparently as a result of the electron falling into the hole. In the modern perspective, an electron encountering a negative energy hole looks exactly like annihilation with its antimatter positron partner.

Despite implications that seemed ripped from the pages of a pulp science fiction book, the modern version of Dirac's theory, quantum electrodynamics, is recognized as one of the greatest advances in physics of the modern age. It melds relativity, quantum mechanics, and electromagnetism to describe the properties and interactions of the smallest building blocks of matter with extraordinary accuracy and precision. It was what legendary physicist Richard Feynman called the "Jewel of Physics."[4]

As it happens, the abbreviation for quantum electrodynamics, QED, also stands for the Latin phrase *quod erat demonstrandum* meaning literally "what was to be demonstrated." The letters are traditionally appended to logical and mathematical arguments to indicate a problem has been fully and finally solved and that the calculations have come to an end. QED is not the end of physics, of course, but instead an important step in the development of quantum field theories that have transformed the way scientists understand particle physics and the cosmos.

Draining the Dirac Sea

While he acknowledged that Dirac's development of QED results in a working theory, Majorana pointed out in a paper he wrote in 1933, and ultimately published in 1937,[5] that the premises of Dirac's tour de force derivation were inelegant. Majorana set out to offer an alternative derivation of Dirac's QED that led to the same predictions, while avoiding the strange negative energy sea that Dirac's approach required. Along the way, although it was not his stated goal in the process of reconstructing QED, Majorana discovered a model of the neutrino that looms large in modern physics theory and experiment. Twenty-three years before its experimental discovery, Majorana showed that the most simple and elegant model of the neutrino suggests that it is its own antiparticle.

Majorana's approach, and the insights his new derivation of quantum electrodynamics provided, echo many revolutions in science and mathematics over the millennia, including one of the most scientifically transformative periods in history: the development of the sun-centered model of the solar system. When Ptolemy set out, almost 2,000 years ago, to describe the motions of the sun and planets in the sky, he began with a premise that seemed reasonable at the time. He assumed that humanity and, by extension, the Earth, was at the center of the universe. It was not a bad guess to begin with. While his perspective is often attributed to mystical beliefs that humans are special and therefore must be at the center of creation, it was largely based on the observation that the stars appeared to be fixed in the sky relative to the Earth. With the observation methods and tools available at the time, early astronomers saw no indication that we were moving through the universe.

Ptolemy's model was surprisingly accurate and precise, but it demanded complicated motions of the planets to make predictions. That in turn implied a complex celestial machine that guided the heavenly bodies on intricate, looping paths. Ptolemy proposed that sun and planets were adhered to enormous spheres that made up a clockwork mechanism with the Earth at the center. The spheres could not be detected in any way, except for their control over planetary and solar motion.

The Renaissance astronomer and mathematician Copernicus realized that modeling the solar system with the sun at the center offered a much more intuitive framework that could provide comparable predictions to Ptolemy's system[6]—except without mystical spheres.

We know now that neither the Earth nor the sun are at the center of the universe, or even at the center of our galaxy. The simplicity of the sun-centered model of the solar system, though, provides an example of a powerful conceptual view that we can apply to the motions of any collection of planets and stars. It's far superior to Ptolemy's geocentric model, which paints even our comparatively simple solar system in complicated terms, and would be almost inconceivably difficult to apply to anything beyond the sun and the nearby planets. Ptolemy's model of the planets and sun, though practically useful and mathematically true enough, lacked the symmetry and simplicity to enable and inspire broader scientific progress. The elegant heliocentric model of Copernicus, by contrast, sparked a revolution that transformed the sciences over the course of a few hundred years.

Majorana recognized the sorts of issues that plagued Ptolemy's model of the solar system in Dirac's derivation of quantum

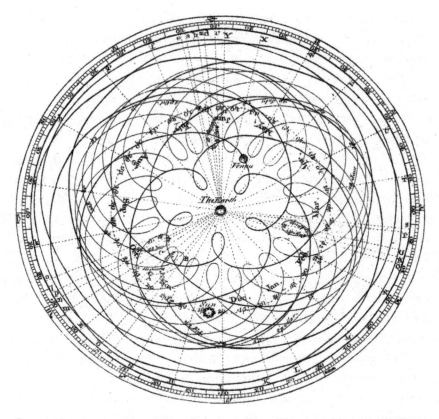

Geocentric representation of the apparent motion of the sun, Mercury, and Venus, from the "Astronomy" article in the first edition of *Encyclopaedia Britannica* (1771). This diagram shows the sun's apparent annual orbit, the orbit of Mercury, and the orbit of Venus as viewed from the Earth. *Source: Encyclopaedia Britannica* (1st edition, 1771; facsimile reprint 1971), vol. 1, fig. 2 of plate XL facing p. 449.

electrodynamics. In both cases, the derivations begin with reasonable assumptions that require contrived models and complicated math, on the way to creating mathematically correct and powerful theories.

In Ptolemy's case, the difficulty arises with his assumption that the Earth is special, treating it asymmetrically with respect to all the other planets. In Dirac's case, the premises he begins with treat the electron asymmetrically with regard to its antiparticle, the positron.

Majorana titled his groundbreaking paper "A Symmetric Theory of Electrons and Positrons," making it clear that his intention was to eliminate the contrivances and difficulties that Dirac's approach created. As was the case with the sun-centered replacement of the geocentric solar system model, Majorana's symmetric derivation of QED provided immediate insights, plus some additional benefits that are driving major experimental efforts today.

Majorana notes in the very first line of his paper that QED is a good theory of the electron and its antiparticle, despite the fraught initial assumptions. "The interpretation of the so-called 'negative energy states' proposed by Dirac leads, as it is well known, to a substantially symmetric description of electrons and positrons," writes Majorana, before pointing out the inherent troubles: "The prescriptions needed to cast the theory into a symmetric form, in conformity with its content, are however not entirely satisfactory . . . because one always starts from an asymmetric form and because symmetric results are obtained only after one applies appropriate procedures," in order to handle infinite values that plague the derivation, Majorana argues. "For these reasons, we have attempted a new approach, which leads more directly to the desired result."

Fundamentally, there's no reason to treat electrons and positrons differently, other than the fact that the existence of electrons was established decades before Dirac tackled the problem, while positrons had not yet been discovered when he first developed QED. The electron and positron are complementary partners, with the same mass but opposite charge. Therefore, Majorana argued, they deserve equal treatment in theories that describe them.

Just as Ptolemy made a reasonable, but confounding, assumption that our place in the universe is special based on the observations and information he had available to him, Dirac had treated the electron asymmetrically with respect to the positron simply because no one knew about the electron's antimatter partner at the time.

As the history of solar system models shows, however, there's more to a theory than providing true and useful information and predictions. There's also the potential to offer insight that leads to new discoveries and inspiration for subsequent advances in theory. Dirac's negative energy sea was tantamount to the spheres that guided the motions of the planets in Ptolemy's solar model: invisible, inscrutable, and undetectable in any way except for the effects they are proposed to explain.

In both cases the bizarre underlying mechanisms were part of seemingly circular arguments that obscured nature rather than illuminating it: The planets move across the sky because they are attached to a system of invisible spheres, and the invisible spheres must exist because of the way the planets move; the negative energy sea must exist to prevent electrons from emitting infinite amounts of energy as they fall to infinitely negative energies, and the fact that electrons don't emit infinite amounts of energy means the negative energy sea must exist.

The need for negative energy electrons in QED, as far as Majorana was concerned, was a symptom of the awkward assumptions that went into developing the theory in the first place, as was the case for Ptolemy's planet-guiding spheres. Majorana notes that, practically speaking, his derivation doesn't change QED predictions as far as charged particles like electrons and positrons are concerned, in the same way the Copernicus model didn't make any predictions about planetary motions that were at odds with predictions in Ptolemy's model. Nevertheless, by treating the electron and positron on the same footing, Majorana eliminated the disquieting notions that Dirac's derivation required.

"In the case of electrons and positrons, we may anticipate only a formal progress; but we consider it important, for possible extensions by analogy, that the very notion of negative energy states can be avoided," wrote Majorana as he essentially erased the need for the negative energy sea from QED at all. "We shall see, in fact, that it is perfectly, and most naturally, possible to formulate a theory of elementary neutral particles which do not have negative (energy) states." Later in the paper, Majorana derives an extraordinarily concise set of equations that applies to particles that have mass but no electric charge.

At the time, there were two uncharged particles that Majorana thought the theory might apply to: the neutron, which was discovered the year before Majorana wrote his paper, and the neutrino, which Pauli had proposed in 1930 but had yet to be discovered.

"The advantage," wrote Majorana in explaining the significance of the equations, ". . . is that there is now no need to assume the existence of antineutrons or antineutrinos." That is, the equations suggested that neutrons and neutrinos are blends of two states: one state that we consider the particle and the other that we consider the antiparticle.

As an illustration, imagine a child on a swing. When the child is at the maximum forward swing position, they are in one state, and at the maximum reverse swing, in another state. At a certain instant, the child may be 10 percent away from the forward state and 90 percent away from their reverse state. At a later time, the blend of states will be switched. And sometimes they will be fully in the forward state or fully in the reverse state. No matter what, it's the same child on the swing.

If Majorana's model holds, then the same could be true of some neutral particles. Such a particle would be capable of fulfilling roles typically attributed to both their matter and antimatter versions in experiment, with the likelihood of it behaving like one or the other depending on the blend of the two states. If so, it's meaningless to distinguish between the particles and their antimatter partners.

In the quantum world, we can't directly see the child's swing partway between the two states. Instead, experiments involving a quantum mechanical analog of a swing would reveal one state or the other with different probabilities. A quantum swing that's 90 percent forward looks like a fully forward swing 90 percent of the time, but like a fully backward swing 10 percent of the time. That is, swing positions (or the matter/antimatter behavior of a neutral particle as Majorana envisioned it) are quantized states of a quantum mechanical system. While it is built of both, it manifests as only one in experiments and interactions. It's not until we make a measurement that we can know which of the two quantum states it ends up in.

In Majorana's approach to QED, the same would be true of a neutral particle that is its own antimatter partner. At any moment it's a blend of both matter and antimatter versions, but it will turn out to be one or the other in an experiment, with the odds depending on where it is in its swing between the two states.

Majorana was mistaken with regard to neutrons, likely due to the fact that they had appeared in experiments only a few months before he wrote his paper. Unlike neutrinos, neutrons are composed of quarks, each of which carries an electrical charge and an additional quantity called the baryon number. As a result, quarks can't be their own antiparticles. Because neutrons are built of quarks, they can't be their own antiparticles either. However, the issue is still open with regard to neutrinos, which are elementary particles with no internal structure. It's possible, and many physicists believe likely, that neutrinos are their own antiparticles.

In honor of Majorana's insight, the neutrinos that fit the model he discovered in reformulating QED are now called Majorana neutrinos. Dirac neutrinos would have distinct antimatter partners. No measurement has yet been performed that can definitively say whether the neutrinos we observe are of the Dirac or Majorana variety. Experiments currently underway, and others in development, could settle the question soon. They may ultimately show that neutrinos are as Dirac envisioned, but the simplicity and symmetry of Majorana's version of QED is compelling. To find that the particles are Dirac neutrinos rather than Majorana neutrinos would be essentially comparable to discovering that the Earth truly is, in some way, at the center of the universe—a situation that's possible, and also unsatisfying, inelegant, and improbable.

The Mysterious Disappearance of Ettore Majorana

Majorana's paper inspiring the quest for evidence of the symmetry between matter and antimatter neutrinos was his final scholarly work. It's also a key exhibit in the tumultuous months and days that were likely Majorana's last.

Beginning in 1928, he had been among Enrico Fermi's most promising students. Along with Bruno Pontecorvo, Majorana was one of a group of physicists under Fermi's wing known as the Via Panisperna Boys for the street in Rome that was home to the Physics Institute of the Sapienza University. Due to his tendency to meticulously interrogate the work of his peers and mentors alike, Majorana was considered the Grand Inquisitor of the group. He could be overheard from time to time in heated discussions even with Fermi, who was referred to as The Pope of the Panisperna Boys.[7] His time at the University in Rome was the most fruitful period of Majorana's career. It was also brief.

Following a trip to Germany in 1933 to visit one of the leading architects of quantum mechanics, Werner Heisenberg, Majorana fell ill. Intestinal distress, probably due to an ulcer, possibly in combination with what his family and acquaintances described as intense shyness, led Majorana to retreat to his sister's home in Rome. He stopped going to the university and became something of a hermit, with occasional convalescent trips to the shores of Croatia on the Adriatic Sea.

Though never a prolific author, Majorana's publication output had ceased entirely. His sister would later recall Majorana working late into the night on what she believed were theoretical studies.

Other than a line or two in some letters, no papers, notes, or documents of any kind have come to light to support her perceptions.

Majorana dutifully responded to letters from his uncle Quirino, an experimental physicist, checking equations from time to time and offering gentle feedback on some theoretical issues. As far as anyone can definitively tell, the superficial and sparse responses were the full extent of his scientific activity during his seclusion.

In 1937, Enrico Fermi was determined to pull his brilliant protégé back into the academic sphere. The opportunity Fermi needed to draw Majorana out arose with a competition to fill several professorships in Italy. The selection committee was composed of five distinguished Italian physicists. Fermi, as the most accomplished member as well as being the chair of the committee, was ideally situated to place Majorana in a position at one of the nation's leading institutions. It only made sense that a disciple Fermi described as comparable in brilliance to Galileo and Newton should take his place in one of the great Italian universities that were standing at the forefront of physics at the time.

But Fermi had a problem: Majorana's academic inactivity and barren publication list meant that he couldn't be on par with several other applicants, when compared under the competition guidelines. Fermi managed to partially address the issue by pressuring Majorana to submit for publication his paper on the symmetric theory of the electron and the positron, which had languished without publication since at least 1933. The paper that would later become Majorana's greatest legacy was a crucial element of Fermi's plan to hand Majorana a university position. Still, there was the matter of Majorana's lengthy absence from academia.

The solution that Fermi and the commission came to was this: If Majorana couldn't measure up to the other applicants on the competition guidelines, it would be better not to judge him by those standards at all. In a letter to the Italian Minister of National Education, they argued that Majorana was in such high regard in the physics community worldwide that "the Commission hesitates to apply to him the normal procedure for university competitions."

The commission also coyly noted, in an addendum addressing Majorana's publications, that while his work was "quite remarkable for the originality of the methods employed and the importance of results achieved," they would not include the full list of his papers. In lieu of what would have been a thin and dated publication list, they tallied only the most important three. It was a wise choice, considering that the sum of Majorana's publication output consisted of a scant nine papers.

Other leading contenders had much more extensive publication lists. During Majorana's five-year doldrums, Gian Carlo Wick authored at least fifteen, and Giulio Racah published eleven or more. Wick would win a professorship at the University of Palermo, and Racah would go to the university in Pisa, both based on the merits as evaluated through the prescribed competition procedures.

There is no question that the three highlighted Majorana papers were outstanding. In fact, essentially all Majorana's publications are remarkable. That didn't stop Fermi and the commission from employing a final bit of subterfuge. In summarizing Majorana's work on the symmetrical description of electrons and positrons, they described it as "a recent paper." While technically true, as it was the last of Majorana's papers and was published in *Nuovo Cimento* a few months before the competition, it gave the false impression that he was actively involved on the forefront of theoretical physics. As exceptional and original as it still was in 1937, the paper was already over four years old and no reflection of Majorana's recent activity—or, more precisely, of his inactivity at the time of the competition. The ploy was a success and Majorana was appointed a professor at the University of Naples.

Based on a letter he sent to his physicist uncle, Majorana had not been involved in, or even aware of, Fermi's scheme prior to the assignment of his position. "I laughed a little at the procedural oddities surrounding my competition," he wrote to Quirino, "of which I had no idea."

Majorana also made light of the Fermi-orchestrated fix in a letter to his close friend, Giovannino Gentile, who had won a professorship through the competition, but without any abnormal intervention. After congratulating Gentile on his new position at the University of Milan, Majorana acknowledges the special treatment that Fermi had arranged with a jest. "Pius XI is very old," Majorana wrote, "and I received an excellent Christian education; if at the next conclave they make me Pope for exceptional merits, I will without doubt accept."

By November, Majorana was corresponding about the appointment details with his future boss Antonio Carrelli, the director of the physics institute at Naples. Carrelli had also served as a member and secretary of the commission Fermi chaired. On January 12, 1938, Majorana formally accepted his position as theoretical physicist at the University of Naples. He presented his first lecture at the university the next day.

A little more than two months later, Majorana disappeared entirely. Neither his family nor his colleagues ever heard from him again.

Case Closed

Seventy-seven years after the brilliant young physicist suddenly vanished, Italian prosecutor Pierfilippo Laviani declared the Majorana mystery solved. The legal opinion is unlikely to put an end to the enduring debate about what really happened to Ettore Majorana, whose brief career and short publication list had a disproportionately large influence on particle physics.

The facts in Majorana's mystery are few. The following are the events in the final two days before he disappeared:

March 25, 1938

- Majorana writes a letter to Director Carrelli of the Naples Physics Institute announcing that he has made an unavoidable decision. He's not specific or clear about what the decision is, but he apologizes for his sudden disappearance and asks the director to remember him to his colleagues at the institute.

- He writes a note to his family and leaves it on the desk in the hotel room where he last stayed in Naples. It reads:

  To my family [on the outer envelope]

  Naples, 25th March 1938-XVI—I have just one wish: that you do not wear black. If you wish to mourn me then do so, but not for more than three days. Afterwards, if you can, keep my memory in your hearts and forgive me—affectionately, Ettore.

- Majorana empties his bank account of the few months of salary it holds and boards a steamship to Palermo.

March 26, 1938

- On arrival in Palermo, Majorana sends a telegram to the institute director rescinding his previous letter. He then writes a second letter to the director promising to return but affirming his plan to resign his teaching post.

- He buys a seat on a boat sailing from Palermo to Naples that evening.

- Majorana disappears, probably after boarding the boat and prior to arriving in the port at Naples.

The circumstances of Majorana's disappearance are widely argued. Multiple theories have been proposed, including his suicide, murder, kidnapping, escape to a monastery, or even that Majorana resigned himself to the life of a street beggar. For 70 years, there was little beyond speculation and armchair detecting to add to the story. Then, in 2008, during an Italian television show dedicated to finding missing persons, a caller to the program claimed to have some clues.

On air, Roberto Fasani said that in 1955 he had met a man in Caracas, Venezuela, who went by the name Mr. Bini. Fasani recalled someone telling him Bini's real name was Majorana. He also claimed to have discovered a postcard in Mr. Bini's car dated 1920 and addressed from Ettore Majorana's uncle Quirino Majorana to an American physicist.

When the case went to a hearing, Fasani produced a photograph of Mr. Bini that showed him to be about the age Majorana would have been in 1955 and having a facial bone structure similar to Majorana's father, according to expert testimony. It was enough, Prosecutor Laviani pronounced, to prove that Majorana had cut all his personal and professional ties, fled to Venezuela, and lived anonymously of his own free will for at least 17 years. Therefore, Laviani concluded, there was no sign of a crime and no need to keep the case open.[8]

Many, including Majorana's most dedicated biographers, find Laviani's legal decision an unconvincing one. For anyone interested in studying the Majorana story, the rabbit hole is deep and convoluted—it's a poignant capstone to the enigmatic life of Ettore Majorana, whose insight into the neutrino still powerfully resonates today.

The simplest answer is not always the right one, but it's not a bad first guess. In the case of Majorana's disappearance, his own writings and evidence of his final movements suggest that anyone wielding Occam's razor would likely come to at least the initial conclusion that the solution to the mystery is indeed simple: Majorana leapt to his death from a boat traveling between Palermo and Naples on March 26, 1938. If so, we lost to the Tyrrhenian Sea the very person who strove to save physics from Dirac's infinite, negative energy particle sea.

5

Turn On the Sun

"How often have I said to you that when you have eliminated the impossible, whatever remains, however improbable, must be the truth?"

—Sherlock Holmes, *The Sign of the Four*[1]

Something was wrong with the sun.

Perhaps the core was rapidly rotating, leading to lower pressures and reduced rates of fusion. Maybe it wasn't running on fusion at all, but instead was powered by a black hole at its center that sucked in matter and spewed out energy. Worst of all was the possibility that it was at the end of its supply of fuel. It might be that the hydrogen that had sustained it for four and a half billion years was exhausted and it was beginning its inevitable decline to stellar death.

Regardless of the cause, the signs were clear: A fraction of the neutrinos that should be emitted from the sun's core were making it out.

Over 34 years, beginning in 1967, Ray Davis of Brookhaven National Laboratory had checked and rechecked the experiment he was leading deep in an active gold mine in South Dakota. It was working, counting neutrinos as it should. There just weren't nearly as many as his theoretician partner, John Bahcall, calculated that a star like the sun produces. The rate was about a third of what Bahcall believed that established theory predicted.

Stars are substantially transparent to neutrinos, which means the ones that Davis was finding provided a snapshot of the sun's fusion reactions from a little over eight minutes prior, a delay due only to the time it takes for a neutrino to travel to Earth.

The heat and light the sun sends out, however, was generated hundreds of thousands of years ago in the solar core, then diffused outward until reaching the corona where it emerges into space. If reactions driving the sun were to wink out like a candle in the wind, we could see the change in the neutrino output almost immediately, but the sunlight would shine on steadily for about 170,000 years before there'd be any outward sign of trouble.

Exactly what was going on with the sun was a mystery, but the discrepancy between the expected neutrino flux and Davis's measurements pointed to something profound happening, whether in the sun itself, in our understanding of neutrinos, or in our knowledge of the fusion processes that create them.

Going Underground to Study the Sun

Davis's failure to find neutrinos in his Savannah experiment had been a clear indication that there were two types of neutrinos: the matter neutrino capable of converting chlorine to argon, and its antimatter partner, which was emitted in fusion reactors but that couldn't interact with chlorine. The neutrinos that should have been streaming out of the sun, on the other hand, were the right type to turn up in the chlorine-to-argon experiment. Although the reaction Davis relied on was not good at catching reactor neutrinos, the null result suggested it could be ideal for detecting neutrinos from the sun.

Most solar neutrinos, which come from the fusion of hydrogen atoms to form helium, are too low in energy to turn up in Davis's experiment. One in 10,000 of the neutrinos from stars like ours emerge with energies high enough to drive the chlorine reactions Davis was after. Based on the efficiency he calculated for capturing them in the dry-cleaning solution, Bahcall estimated that Davis would find between four and eleven neutrinos per day of exposure to solar neutrinos, provided he used a 100,000-gallon tank of the fluid.

To reduce the false positives that would come from cosmic rays, Davis began building the Brookhaven-backed experiment in 1965 deep in in the Homestake Gold Mine in Lead,[2] South Dakota. The chamber where the experiment would reside was excavated 1,478 meters underground. The tank was far too large to move through mine shafts, so welders assembled it piece by piece in the underground chamber.

Once the tank was completed and filled, the experimenters would extract radioactive argon from the tank every few months with an intricate series of steps that involved bubbling hydrogen through the dry-cleaning fluid to dissolve and carry out the argon. They would then supercool the hydrogen-argon mixture and catch the argon in a cold charcoal trap. The minute amounts recovered were placed in counters where Davis could determine the number of radioactive argon atoms by looking for the X-rays they gave off as they decayed and turned back into chlorine.

In order to measure the amount of accumulated argon with as little background interference as possible, Davis's team inserted the argon-filled counters into repurposed sections of battleship gun barrels that had been forged before the advent of the atomic bomb. Steel produced after the development of the bomb picks up atmospheric contamination released from nuclear weapons tests.

Although such impurities are minuscule, they would be trouble-some for an experiment that collects a handful of atoms a day. The low-background steel, as metal forged before the atomic age is known, offered uncontaminated shielding that added to the protection that the kilometer and a half of rock and earth above the experimental chambers provided.

Despite all their efforts, it was clear from even their preliminary measurements that the number of solar neutrinos was coming up short. Davis's concerns were apparent in an update he sent to theoretician William Fowler in late 1967: "We are ready now," wrote Davis, "turn on the sun."

Instead of counting an average of seven and a half neutrinos per day, as Bahcall had calculated they should, Davis was finding two and a half. Neither the joking plea to Fowler nor any refinements to Bahcall's math could bring the expectations and observations into line. The mysterious incongruity persisted.

The Mystery Deepens

The leading suspects in the seeming discrepancy between Bahcall's calculations and Davis's experiment were the detector in the Homestake mine, theoretical models of the stellar fusion, or a problem with neutrinos themselves. Davis primarily distrusted stellar fusion theories. But as an impeccable experimentalist, he set about ensuring the detector wasn't to blame, just in case.

Davis and his team came up with a variety of procedures to ensure their argon-capturing methods worked: They introduced radioactive sources to the detector to proactively create argon atoms from chlorine and confirm that they could find them with the extraction system; they injected radioactively labeled chlorine compounds to produce a different form of argon and recovered the amount they expected; and they looked for any of the non-radioactive forms of argon that might have seeped in from the outside air in order to confirm the tank welds were impervious. In one of the most extraordinary tests of the system, the team placed 500 radioactive argon atoms into the 100,000 gallon tank, without telling Davis the number in advance, and challenged him to count them. Davis found them all.

To put it in perspective, imagine trying to find a single, specific grain of sand hidden amid all the dunes on a beach. That would be child's play compared to the hunt for argon in the Homestake mine experiment. Davis was digging for atoms in a tank that contained trillions of times as many molecules as there are grains of sand on all the beaches of the world put together.

In the decades that he searched for neutrinos in the Homestake mine, Davis identified no flaws in the detector or their methods. The result seemed inescapable: The neutrino count was correct. That left the fault to either our understanding of the sun or to neutrinos.

The range of possibilities related to the sun extended from the merely unlikely to the fantastical: a black hole at its center, radical modifications to established quantum mechanics, or the hair-raising death of the star that drives the engine of life on our planet, to name a few of the more extreme implications of the solar neutrino problem. In 1968, Pontecorvo, who had proposed the reaction at the core of Davis's experiment in the first place, advocated an elegant alternative that pointed the finger at the neutrinos as the most likely suspects.

Pontecorvo had moved to England after his stint in Canada. In late 1950, he and his family suddenly and mysteriously disappeared. According to CIA reports, shortly after defecting he was leading efforts to refine uranium for nuclear power plants.[3] Newspapers at the time suggested Pontecorvo was lending his expertise to help the Soviets with their atomic weapons program. "[I]t is reported," read an article in the April 25, 1953, edition of the *Svenska Dag-bladet* daily newspaper in Stockholm,[4] "that work on the Russian hydrogen bomb has progressed so far that practical research with an experimental bomb will be conducted in July of this year under the direction of Professor Pontecorvo." The news article was wrong by one month. The test would take place in August of 1953, not July.

By 1956 Pontecorvo resumed publishing scientific papers from his new homeland, the Soviet Union. Once again, his interests turned to neutrinos.

With prominent Russian theorist Vladimir Gribov, Pontecorvo proposed that theories of the sun's fusion cycle were correct, meaning both that electron neutrinos flowed out from the solar reactions, just as expected, and that Davis's experimental design was effective at catching them. Fewer electron neutrinos were arriving in the Homestake Mine detector than expected not because there were fewer coming from reactions in the sun, Gribov and Pontecorvo argued, but because the neutrinos were changing their identities on route to the Earth.[5]

Much like Davis's experience with the Savannah River reactor experiment, the fact that neutrinos were missing was almost as important as finding them would have been. In Savannah River, it established the existence of antineutrinos. In Lead, South Dakota, it was a sign something much more profound was going on.

In 1959 Pontecorvo had first proposed that the neutrino associated with the electron's heavier cousin, the muon, was in fact different from the electron's neutrino.[6] Pontecorvo and Gribov showed that the transformation from one neutrino variety to another could occur as a result of the mass difference between the electron and muon neutrino varieties. Considering that it wasn't clear at the time that neutrinos had any mass at all, the theory was not initially a leading candidate to explain the solar neutrino problem. The landscape changed three years later when Lederman, Schwartz, and Steinberger, though unaware of Pontecorvo's proposal, experimentally confirmed the distinction between electron neutrinos and muon neutrinos.

If there were more than one type of neutrino, then, as Pontecorvo and Gribov suggested, neutrinos emerging from the sun could oscillate from one variety to another as they traveled 150 million kilometers through space to the Earth. They would arrive as a mixture, instead of the purely electron neutrino population that should have been created in stellar fusion. The neutrino varieties are now referred to as *flavors*. At the time that Pontecorvo and Gribov proposed that neutrinos shift from one flavor to another, they concerned themselves only with the electron and muon flavors.

In the mid-1970s, another, yet heavier cousin of the electron, called the tau lepton, turned up. If the electron and muon come with associated neutrinos, then the tau should too. In that case, the tau neutrino is involved in oscillations as well, potentially complicating the flavor shifting further.

Assuming neutrinos do oscillate among varieties, Lincoln Wolfenstein of Carnegie Mellon University found that passing through matter should change the oscillations. Soviet physicists Stanislav Mikheyev and Alexei Smirnov calculated that after neutrinos were created in the sun, oscillations would be enhanced as they moved from the core and through the solar plasma before escaping.

Although fusion neutrinos pass easily through the sun, the MSW effect (which takes its initials from the last names of Mikheyev, Smirnov, and Wolfenstein) shows that stars are not entirely transparent to neutrinos. It's comparable to the way that light can shine through a fish tank but is affected by the water along the way. Just as light slows in water, neutrinos slow during their passage through the dense plasma of the sun, according to the MSW theory. This gives them more time to oscillate and increases the chances that electron neutrinos born in solar fusion are a different variety by the time they emerge.

Mystery Solved

Support was growing for the idea that neutrino oscillations may be the cause of the deficit in Davis's measurements. Harvard physicist Sheldon Glashow, though, was among the holdouts who suspected the sun might be to blame rather than the particles. His view was less apocalyptic than some.

Our star was not dying. He supposed, instead, the shortage of neutrinos was just a phase the sun was going through as the result of seismic fluctuations in the solar core. Perhaps, he and coauthor Alvaro De Rújula of CERN advocated in 1992,[7] it was a matter of being patient and collecting more neutrino data over time. We could be at an ebb in neutrino production resulting from solar pulsations. In that case, neutrinos would be the solar equivalent of seismometers on Earth, providing a readout of helioseismic activity.

"If neutrino experiments were to detect the sun's heartbeat," wrote De Rújula and Glashow, "it is the sun that oscillates, not the neutrino."

As Glashow recalled 30 years later, "There was some doubt about neutrino oscillations at that time," just before a definitive announcement that broadly settled the issue.[8]

While he was among those who harbored doubts about the oscillation explanation for the solar neutrino deficit at the end of the last century, Glashow was an early proponent of looking for signs of oscillation in neutrinos created when cosmic rays smash into the atmosphere. He proposed the search in a talk he gave at a September 1979 "Quarks and Leptons" conference in France.[9] Reines pitched the same experiment in more detail a month later at the 16th International Cosmic Ray Conference in Kyoto, Japan.[10]

Both Americans, however, were well behind neutrino-whisperer Pontecorvo, who was already leading a search for atmospheric neutrino oscillation. He and his colleagues at the Neutrino Observatory of the Institute for Nuclear Research of the Academy of Sciences had described the experiment in 1976. By the following year, they were building it at the underground facility near Moscow. Pontecorvo calculated that it was at least marginally capable of detecting atmospheric neutrino oscillations.[11]

Despite their early start, the Soviets would not win the race to find neutrino oscillations. Instead, Japan's flagship Super-Kamiokande neutrino detector was the first to see the effect in atmospheric neutrinos, with the Canadian Sudbury Neutrino Observatory confirming solar neutrino oscillations three years later.

"I was at the conference in 1998 in Takayama, Japan when the Japanese made their announcement of the observation of atmospheric neutrino oscillations," says Glashow, referring to the revelation that fewer muon neutrinos were coming up through the earth than raining down from above. It was an indication that they must be changing identity as they passed through the planet. "That was the point where everyone began to believe . . . that there were indeed neutrino oscillations." The response to the announcement, as Glashow recalls, was unusual in at least one respect. "Never before did I see a standing ovation [at a science conference]. When they announced their decisive evidence for neutrino oscillations, the audience rose to a man and a woman, and applauded," said Glashow in a 2019 interview. "It was fantastic."

The third flavor, the tau neutrino, was finally detected in 2000 at Fermilab, outside Chicago. In 2001, a detector that could see all three neutrinos, the Sudbury Neutrino Observatory (SNO) in northern Ontario, confirmed that the sun emitted neutrinos in just about the quantities Bahcall had calculated. In combination with the Homestake Mine measurements, it was clear that neutrinos were oscillating among the flavors, as Pontecorvo had suspected.

The missing particles in Davis's experiments were the first to offer a glimpse into the mysteries surrounding neutrino oscillations. The combination of Davis's results with those of SNO, however, still allowed for several possible values for the neutrino properties that governed the oscillations. Definitive data came not from further observation of the solar neutrinos themselves, but from a reactor experiment in Japan.

The KamLAND detector was installed in a mine in Japan in 2002, a location previously occupied by the Kamioka Neutrino Detector (Kamiokande). It is composed of a sphere of liquid scintillator, surrounded by photomultiplier tubes to detect the flashes of light generated when a neutrino interacts inside the scintillator. The surrounding countryside is home to more than 25 reactors at an average distance of 180 kilometers, which were producing antineutrinos with an energy spectrum in the vicinity of the solar neutrinos that Davis had been studying. This gave KamLAND access to the same range of oscillation parameters that governed solar neutrinos.

The first results from KamLAND, released in late 2002, removed the previous ambiguities hanging over the solar neutrino problem, bringing the long journey begun in the Homestake Mine almost four decades earlier to a successful conclusion. KamLAND was able to measure the neutrino oscillations to show how they slow as energy increases, as theory predicts.

More recently, detectors including Borexino in Italy have expanded the study of neutrinos to low-energy particles coming from other phases of solar fusion, allowing the potential comparison of oscillations over the same distance that Davis and Bahcall studied with the Homestake Mine experiments, but at different energies.

Earth-based experiments offer measurements over shorter distances, ranging from roughly the diameter of the planet down to tens of meters. Even before the definitive discovery of oscillations with the Sudbury Neutrino Observatory in 2001, the Super-Kamiokande experiment in Japan found differences in the number of neutrinos produced by cosmic rays striking the atmosphere. Twice as many neutrinos were arriving from cosmic rays hitting the atmosphere above the detector as from the cosmic rays impinging on the atmosphere on the other side of the planet, suggesting that neutrinos were transforming into other varieties as they passed through the Earth.

Oscillation studies spanning distances ranging from meters to millions of kilometers are a testament to the fact that neutrinos are more interesting than anyone[12] could have imagined when the particles first turned up missing in Davis's experiments.

Perhaps more important than what the experiments say about neutrinos is their answer to a question that has tormented humanity through the ages. Why does the sun burn? If nuclear fusion is the engine that drives our star, then solar neutrinos must be created in the numbers Bahcall had calculated. Finding not that there was a shortage, but instead that neutrinos oscillate, finally confirmed that the sun runs on nuclear fusion.

"The collision between solar neutrino experiments and the standard solar model has ended in a spectacular way," said Davis as he accepted the Nobel Prize in Physics in 2002, the year after the definitive confirmation of neutrino oscillations; "nothing was wrong with the experiments or the theory; something was wrong with the neutrinos."

One thing that neutrinos can tell us now is that there's nothing wrong with the sun. It's doing just fine. If that ever changes, neutrinos will let us know about eight minutes after it does.

6

The Incredible Flavor Changer

As Gregor Samsa awoke one morning from uneasy dreams he found himself transformed in his bed into a gigantic insect.

—Franz Kafka, *The Metamorphosis*[1]

Gregor Samsa was a relatively unremarkable traveling salesman, other than the fact that he had been suddenly and mysteriously transformed into an insect in the very first line of Franz Kafka's book *The Metamorphosis*. Neutrinos are interesting particles for a variety of reasons, but much like Samsa, it's their transformations that distinguish them most of all.

As observations confirmed at the end of the last century, Pauli's little neutral one is a blended entity that shifts over time. It may start out as an electron neutrino, then transforms to a blend of electron, muon, and tau neutrinos. It can eventually turn back into an electron neutrino, but most of the time it's, in part, each one. Unlike Samsa, who eventually dies of starvation and despair after a single transformation into a giant beetle, neutrinos typically cycle through flavor blends perpetually.[2]

It's clearer to look at the change that neutrinos go through over distance, rather than time. Consider the Club Sandwich experiment that first concretely proved that neutrinos exist. If you measure the particles with a detector that's placed close to the reactor, you will be able to deduce that they're produced in the numbers that should result from nuclear fission in the reactor.

If you move your detector progressively farther away, the number of neutrinos will decrease. That's not terribly surprising; after all, the neutrinos spread out as they travel away from the source. In the same way, the farther you are from a lightbulb the dimmer it seems to be because the photons it emits are spreading out in all directions, and fewer of them make it to your eye.

For reactors, there is an added reason the electron neutrino flow appears to ebb as you move away, and then, as you continue, starts to increase again. It's as if you walked away from a lightbulb, and at some point it faded out, only to brighten again as you moved away still farther.

If you keep going, moving your detector progressively farther from the reactor, the fading and brightening of the neutrino signal will continue. That is evidence of neutrino oscillations: The odds of detecting electron neutrinos coming from a reactor, or any other source, oscillates with distance from the source.

In reality, it is a tough experiment to perform. Most neutrino detectors are enormous. It would be no easy matter to measure the oscillation in neutrino flow at progressively increasing distances because moving a tank filled with hundreds of thousands of gallons of fluid is prohibitive. Some experiments measure the neutrinos at multiple, fixed distances, which physicists then compare to the theory that describes the oscillation. Others rely on knowing the type and quantity of neutrinos coming from a source and comparing it to the measured types and quantities some distance away.

As Pontecorvo first proposed, neutrinos aren't actually disappearing and reappearing when you move away from a source. They are in fact transforming into a different type of neutrino—a type that an experiment like the one Cowan and Reines built can't see. Most experiments that measure neutrino oscillations suggest that they are transforming into muon or tau neutrinos, although a few suggest other, exotic things may be going on as well.

The meaning of Samsa's mysterious metamorphosis has been debated in literature classes for decades. Although there are significant mysteries remaining regarding the details of neutrinos and their oscillations as well, the theory describing the phenomenon is clear, and it all comes down to flavor and mass.

The Flavor-Mass Connection

Flavors in food are generally blends of spices. Neutrino flavors are blends too. Instead of parsley, rosemary, and sage, the ingredients that generate flavors in neutrinos are known as mass states. The three mass states that combine to make up neutrino flavors are simply numbered m_1, m_2, and m_3.

Each neutrino flavor is made of a different combination of the three masses. Once blended to make a certain neutrino flavor, the relative proportions of the mass states in a neutrino change over time. It's as if you were at a magical reception and the hors d'oeuvres on your plate evolved from sweet to savory while you mingled. If you didn't care for one, just wait and it will be another as you make your way around the room. If you were to get caught in conversation and missed a chance to nibble at the treat on your plate when it was the flavor you desired, wait a little longer, and it will continue changing until it's once again your favorite blend.

Whether it's magical hors d'oeuvres or a neutrino, the shifting from one type to the other is flavor oscillation. You might imagine that some tidbit that was midway through a flavor oscillation from sweet to savory would taste a bit of both. But if you were at a tea party where the food followed quantum rules, the treat you taste will be exclusively sweet or savory. It's one of the central character-istics of quantum mechanics that things come in distinct values, where it's the electrical charge on an electron, the bits of light that are photons, or the flavor of a neutrino. The chances that you will end up with one or the other flavor changes, but you will never taste some of both at the same time.

You probably won't get invited to a party with flavor-oscillating food, but neutrino flavors oscillate all the time. It's the reason Davis's count of solar neutrinos came up short—he was after a plate of savory electron neutrinos, and he couldn't see the sweet muon or tau neutrinos passing through the room.

How a party treat might change from one flavor to another is fod-der for the imagination, but the rules that govern neutrino flavor oscillation depend on the differences between the three neutrino ingredients. To be more precise, in the equations that describe oscil-lations, it's the difference in the squares of the masses of the three states that matters. We don't need to know the actual masses of the three states to understand how neutrinos oscillate. That's fortunate, because neutrino masses are extremely small and difficult to mea-sure, while oscillations are often hard to miss. As a result, you can measure oscillations to get a handle on the differences between the mass states, even if they can't tell us what the specific masses are.

It's a little like knowing that you need a pinch of paprika, two pinches of salt, and a handful of parsley. It's hard to say exactly how much those quantities are in terms of teaspoons and cups, but when you taste what comes out of the oven you can tell which you added a lot of and which you added a little of.

Observations of neutrinos that travel relatively long distances, like the ones coming from the sun, depend mostly on the difference between two mass states that appear to be relatively close together. For shorter-range observations, such as those involving neutrinos produced in the Earth's atmosphere, the oscillations suggest a large mass-squared difference. That is, of the three mass states that go into a neutrino flavor, two are clearly close together. They are the states physicists call m_1 and m_2. The third one, m_3, is significantly different. It could be larger in mass than either of the other two, or perhaps smaller. Either way, it stands apart from the other two.

Neutrinos arrive at our quantum tea party in one of the three flavors, like pristine canapés on a silver dish. The sun serves up primarily electron neutrinos; cosmic rays hitting the atmosphere cook up muon neutrinos; and super-high-energy tau particles make tau neutrinos when they ram into atoms.

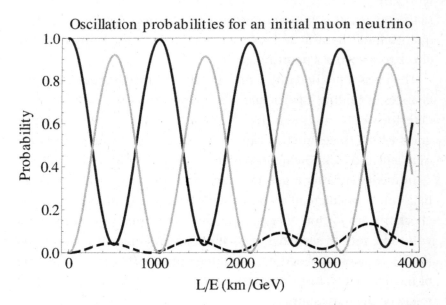

Oscillation probabilities for an initial muon neutrino

Oscillations between three neutrino flavors, where the initial state is a muon neutrino. This is the case relevant to atmospheric neutrinos, and the oscillation is mainly between the muon and tau neutrinos, with only a small amount of electron neutrino present. At increasing distances, the probability that the particle will be found to be one flavor or the other changes due to these oscillations. *Source:* "Oscillations Muon Short.svg," Wikimedia Commons (https://commons.wikimedia.org /wiki/File:Oscillations_muon_short.svg).

Whatever their initial flavor, the neutrinos quickly change to blends of the flavors. It takes time to happen. If you check to see what flavor a neutrino is immediately after it's created as an electron flavor, it will most likely still be an electron neutrino. If you wait longer, as happens if you are checking neutrinos that have traveled to the Earth from the sun, the mass states will have shifted to the point that you have only about a one in three chance of discovering it to be an electron neutrino, and a two in three chance of finding it to be one of the other flavors.

In quantum mechanics, this is known as a superposition of flavors. The mass states are combined in a way that doesn't match with any particular one of the neutrino flavors, making it essentially a blend (i.e., a superposition) of all three. But because it's a quantum mechanical system, we can find only one flavor when we check to see what a neutrino is.

The mass states determine how the flavors are combined, and the chances of finding a particular flavor when you see it in a neutrino detector. You can also go the other way. Just as a connoisseur can guess which spices are in a canapé, we can tell what mass states are in a neutrino by knowing its flavor.

Immediately after a neutrino comes out of a reaction, before it's had time to oscillate, it's a pure flavor. No one can unbake a cake to check the proportion of eggs, flour, and butter that went into it, but you can check to see what mass states are in a neutrino, at least in theory. It takes an experiment that measures neutrino mass, instead of flavor. That is, instead of tasting it to see the flavor, you weigh it to check the mass states.

Just as quantum mechanics allows you to measure only one flavor at a time in a neutrino that has oscillated into a superposition of flavors, it allows you to measure only one of the mass states at a time out of the three that a neutrino flavor is made of.

If you had a tiny neutrino scale, and you set a neutrino on it, it will give you the mass that corresponds to one of the mass states

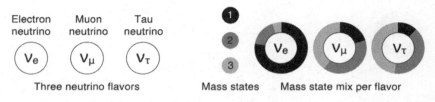

Electron   Muon   Tau
neutrino   neutrino   neutrino

V_e V_μ V_τ

Three neutrino flavors   Mass states   Mass state mix per flavor

Mass blends that make up the pure flavors of electron, muon, and tau neutrinos.
Source: Lucy Reading-Ikkanda for *Scientific American*, July 2020.

(m_1, m_2, or m_3) each time. If you then do the experiment many times with identical neutrinos, you could figure out the proportions of the mass states that make up a neutrino.

A parameter called the mixing angle can tell you the proportion of m_1 and m_2 in the electron flavor (very little of the electron flavor is the third mass state, m_3, so it's okay to ignore it for now). If the mixing angle is about 30 degrees, as experiments suggest, a pure electron neutrino would be about three-quarters m_1 and one-quarter m_2. If you had a collection of electron neutrinos and weighed each of them, in three-quarters of the cases the result would come out to be the mass of m_1, and in the other quarter you would get mass of m_2. A pure muon neutrino is the opposite blend: three-quarters m_1 and one-quarter m_2.

Once a neutrino has been created in a certain flavor state, it immediately begins to oscillate. The higher the neutrino energy, the slower the oscillation. Smaller differences among the mass states masses also slow the oscillations, and the longer a neutrino will have to travel to go through an oscillation. This is why a smaller mass-squared difference involving m_1 and m_2 is important for neutrinos that have traveled a long way, such as those that come from the sun.

A neutrino that starts off as a particular blend and evolves into a different one returns to its original blend after one complete oscillation period. It may start off as pure electron neutrino, but by halfway through an oscillation period, it will have acquired its

Oscillation from an electron neutrino to a tau neutrino. *Source:* Lucy Reading-Ikkanda for *Scientific American*, July 2020.

maximum amount of muon neutrino. In the case of a 30 degree mixing angle, that's about 75 percent. After a full period, it is again 100 percent electron neutrino.

This discussion ignores effects that tend to damp out oscillations. One such factor is that the mass states that make up the flavor state have slightly different energies and therefore travel at slightly different velocities. As time goes on, they separate from each other, which makes it harder and harder for the oscillations to take place.

Speeding Oscillation in the Sun

What we have described are *vacuum oscillations*, because we have ignored any interactions between the neutrinos and their environment. Neutrinos are very weakly interacting, so this is an excellent approximation for neutrinos that travel through the Earth's atmosphere, or indeed through the Earth itself.

The solar neutrinos Davis studied are produced deep in the sun's interior in the presence of a high density of electrons. As neutrinos propagate through the solar medium, their interaction with the electrons reduces their velocity, which leads them to cycle through more oscillations over a shorter distance.

Conditions like these highlight another feature of neutrino flavors and mass states. Just as flavors are made from a combination of mass states, individual mass states can be considered a blend of flavors. In contrast to flavors, mass states don't oscillate. If you somehow had a source of neutrinos all in mass state m_1, then every time you weighed one on your neutrino scale you would find the same reading. If, instead, you ran it through an experiment that measures flavor, you would get electron flavor sometimes and other flavors at other times.

The comparatively high-energy neutrinos Davis studied, in fact, arrived at the solar surface predominantly in a single state of definite mass due to their interactions with the solar material. (If you were to measure their flavors there, you would find about one-third of them to be as electron neutrinos.) Each one is a state of definite mass, and not a blend of masses; no further oscillations take place as the neutrinos travel through the 150 million kilometers from the sun to the Earth. As a result, Davis observed only one-third of the neutrinos that Bahcall's solar model had predicted, not because they were oscillating in flight, but because they had settled into a single mass state that has roughly a one in three chance of registering as a detectable electron neutrino.

Davis's experiment was sensitive to comparatively high energy[3] solar neutrinos. Low-energy solar neutrinos pass through the sun more easily, without interacting much with the material in the sun. They continue to oscillate on their way to the Earth, so the fraction that turn up as electron neutrinos (in detectors that can catch the low-energy neutrinos) is more like 54 percent, instead of roughly 34 percent as it was for Davis's neutrinos.

Additional evidence for the effect that interactions with matter has on neutrino propagation turns up in measurements of electron flavor neutrinos from the sun. Interactions with the matter in the Earth produces a slight enhancement of the electron neutrino fraction during the night, when they also have to pass through the planet on the way to a detector, compared to the daytime, when they travel directly to the detector from the sun. The day/night effect is subtle, but it has been observed in solar neutrino detectors.[4]

Neutrino Mass Hierarchy

Neutrino oscillations give us the crucial information that neutrinos have mass. More precisely, oscillation measurements mean that the three neutrino mass states have different masses. The experiments show that the first and second mass states are close together, and the third is significantly different. Knowing the differences between them doesn't tell us the values of the masses, or which one is heaviest.

This is known as the mass hierarchy question. The ordering is called normal if m_3 is larger than the other two because it would echo the ordering of the heavy tau particle, medium mass muon, and light electron. The electron flavor has more of the m_1 state in it than the other two flavors, and the tau neutrino has the most m_3. The muon neutrino is a more balanced combination of the three.

The reverse of the normal ordering, with m_3 the lightest of the three, is called the inverted ordering. There's no clear reason to expect normal or inverted ordering just yet.

The solar and atmospheric neutrino oscillation measurements don't tell us anything about the neutrino masses directly because they depend only on the relative differences between the squares of the masses. Neutrino mass itself is the target of experiments that look closely at the energy of the electrons coming from the sorts of beta decays that began the neutrino saga in the first place.

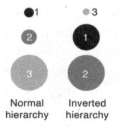

Normal   Inverted
hierarchy   hierarchy

Two possible mass order hierarchies. *Source:* Lucy Reading-Ikkanda for *Scientific American*, July 2020.

The neutrino mass hierarchy can affect the results of other experiments, including those that seek to determine whether Majorana's proposal that neutrinos are their own antiparticles is correct.

Flavor Confusion

It takes only two independent sets of experiments to get a handle on the relative differences between mass states. Solar neutrinos can provide one data set, starting with the Davis experiment that observed a deficit of electron neutrinos coming from the sun. In this case, the baseline distance between the sun and the Earth is very long. Taking into account the interaction of neutrinos inside the sun that enhances oscillations before they escape on their journey to the Earth provides a measure of the effect that m_1 and m_2 masses have on oscillations.[5]

The second data set involves the neutrinos produced by cosmic rays interacting with the upper reaches of the Earth's atmosphere. The most common products in cosmic ray collisions are particles known as pions.

A positively charged pion decays to a muon and a muon neutrino. The muon then decays into a positron, a muon antineutrino, and an electron neutrino. Altogether, twice as many muon-flavored neutrinos emerge from the cosmic ray collisions as electron-flavored neutrinos. Negatively charged pions go through a similar string of decays, also resulting in two muon-flavored neutrinos for each electron-flavored one.

When the cosmic rays first strike the atmosphere, the proportion of muon to electron neutrinos and antineutrinos they create is two to one.

Overall, however, experiments showed their numbers that end up in detectors on Earth are just about equal. In their journey from the atmosphere to the surface, muon neutrinos were oscillating into other flavors, causing some of them to disappear. These are known as atmospheric neutrino oscillations, and they offer an idea of the difference between the third mass state, m_3, and the other two.[6]

It should have been all anyone needed to understand how neutrinos change their flavor blends. But still to be accounted for were data from short baseline distance experiments—those where the source and the detector were sited within the same laboratory.

Sure enough, the results were incompatible with the simple picture of just two independent mass differences. They seemed to show that there must be more mass states involved beyond m_1, m_2, and m_3.

The trouble first showed up in the Liquid Scintillator Neutrino Detector (LSND), which ran at the Los Alamos National Laboratory between 1993 and 1998. In contrast to the solar and atmospheric neutrino experiments, which detected a disappearance of neutrinos, LSND was an appearance experiment.

An intense beam of protons, impinging on a target, generated a beam of almost pure muon antineutrinos, which then traveled 30 meters to the detector. The goal was to look for electrons and neutrons produced when electron antineutrinos appeared and interacted with protons.

LSND researchers found more electron antineutrinos than the solar and atmospheric neutrino experiments suggested. One explanation for the discrepancy could be another neutrino much heavier than the ones that explained the solar and atmospheric data.[7] The picture is that the initial muon antineutrino mixes with the fourth neutrino. The fourth neutrino in turn mixes with the electron antineutrino. The result is a boost in the number of electrons that turn up in the detector.

If this indeed is evidence for a fourth neutrino, it would have to be even more elusive than the other three. This type of neutrino, if it exists, is called *sterile* because it doesn't interact with matter through the weak, strong, or electromagnetic forces. In contrast to the *active* electron, muon, and tau neutrinos, sterile neutrinos can't be directly observed in any laboratory experiment.

It was a disquieting development. No other experiments suggested, or even allowed for, another neutrino beyond the three in the Standard Model.

The LSND result was sufficiently at odds with expectations that a second experiment, MiniBooNE, was mounted to check it. Located at Fermilab in Illinois, MiniBooNE was about 10 times the size of LSND. It was also sensitive to signs of a sterile neutrino. But the design was different enough that it could provide a check on the LSND experiment.

As MiniBooNE prepared to take data in 2002, the smart money, or at least the conservative smart money, was expecting MiniBooNE to refute the LSND result. While LSND was limited to antineutrino events, MiniBooNE detected the appearance of both antineutrino and neutrino electron.[8] MiniBooNE took data from 2002 through 2019 and presented a detailed analysis in 2020.

The surprising conclusion was, all in all, that MiniBooNE agreed with the LSND results. The combined MiniBooNE/LSND data was compatible with a sterile neutrino with a much larger mass than the three active neutrinos.[9] Instead of clearing the air by explaining away sterile neutrinos, the case for them had grown stronger.

A sterile neutrino is not necessarily the explanation of the LSND and MiniBooNE results. But no one has come up with an alternative that's particularly convincing. Other experiments are making compelling cases against existence of a sterile neutrino of the type that LSND and MiniBooNE implied. They include experiments that study oscillations of neutrinos coming from nuclear reactors. The Daya Bay complex in China, in particular, explores several baseline distances. The Minos and Minos+ experiments, based at Fermilab, are using neutrino beams to study neutrinos over longer distances.

These experiments all measure neutrino disappearance, as opposed to the appearance measurements of LSND and MiniBooNE.

But any sterile neutrino should show up either way. The null results from the disappearance experiments make it hard to understand how a sterile neutrino (or perhaps more than one sterile neutrino) could be the explanation of the MiniBooNE/LSND observations.

Given this situation, it is unlikely but not impossible that Mini-BooNE and LSND are just flat-out wrong. If they are right, it's unlikely that a simple fix like one sterile neutrino will suffice to explain their data. Recently, another experiment called Micro-BooNE that uses the same beam line at Fermilab as MiniBooNE has announced its first round of results. MicroBooNE has a more sophisticated detection system, and its initial data see no evidence of the excesses reported by LSND and MiniBooNE. But the earlier LSND/MiniBooNE results still stand.

Other experiments deepen the puzzle still further. In particular, there appeared to be a shortage of electron neutrinos in detectors built of large reservoirs of the metal gallium. The GALLEX (Gallium Experiment) ran from 1991 to 1997 in the Italian Gran Sasso laboratory. It consisted of 101 tons of gallium. The 63-ton SAGE (Soviet-American Gallium Experiment) detector was built in 1989. Both experiments were intended to search for lower-energy neutrinos coming from the sun that would convert gallium atoms to germanium through inverse beta decay.

To calibrate their detectors, the research groups placed neutrino-emitting sources inside their respective gallium reservoirs. They calculated the number of neutrinos they expected to find during the calibration, but the number that actually turned up was about 15 percent short. It's potentially another sign that neutrinos are more complicated than scientists expected, and sterile neutrinos might be the key.

Statistically speaking, the neutrino deficit in gallium experiments is less significant than the LSND/MicroBooNE anomaly. Still, the existence of a sterile neutrino a little heavier than the ones that would explain the other anomalies could be the reason for the missing electron neutrinos in both SAGE and GALLEX.

The sterile neutrino mystery once seemed likely to be nothing more than an error that would evaporate as new experiments came online. Instead, it has only deepened, adding another layer of intrigue to the study of neutrinos.

An End to Oscillations

While Gregor Samsa suffered only a single transformation, neutrinos oscillate through their potential flavors repeatedly, but not endlessly. As a neutrino travels, the mass states that determine how much of each flavor the neutrino comprises move at different rates. It's a direct result of the difference in masses between the states: The less massive the state, the faster it needs to travel to carry a given amount of energy.

Consider, for example, the energy in the motion of a motorcycle and a truck. It takes more energy to move a truck at 100 kilometers an hour than to move the motorcycle at the same speed, a difference reflected in the relative size of the vehicles' engines and fuel tanks. It takes a lot of fuel to move a truck and little to move a motorcycle. Because fuel is the source of energy, its consumption is a measure of the energy involved in getting a vehicle rolling.

On the other hand, suppose you want to use the same amount of energy to move a truck as to move a motorcycle. Enough energy to get the truck barely started will propel a motorcycle to high speeds. Similarly, if an electron neutrino has a certain amount of energy, the lighter mass states that make up the neutrino will move faster than the heaviest one. Over time, the states spread out as the lightest state speeds away and the heaviest one lags behind, until the mass states don't overlap at all.

The separation of states in a quantum mechanical system is an example of decoherence. That is, a coherent blend of states breaks down and the oscillations cease. We're surrounded by neutrinos that stopped oscillating long ago. They're the relic neutrinos left over from the Big Bang and are probably the most abundant type of neutrinos in the cosmos.

With the expansion of the universe, the enormous population of neutrinos that were formed at the very beginning have cooled to the point that they're less than two degrees above absolute zero. Low temperatures mean low energies, and low energies lead to shorter distances over which coherence can be maintained. At a shade under two Kelvin, a neutrino will stop oscillating before traveling even a few tenths of a millimeter. Big bang relic neutrinos have all traveled much farther than that, and no longer oscillate between flavors.

The fact that a neutrino will inevitably end up in a train of separated mass states, rather than a flavor state, reflects the important role the mass states play in neutrino oscillations. They are what physicists call *mass eigenstates*, with properties that do not change with time.

Flavor states are superpositions of the mass eigenstates. They do evolve in time. This is the essential source of neutrino oscillations. But the oscillations will eventually die down. They leave behind the mass eigenstates that are the ultimate end for any neutrino that has traveled past its coherence length since it left the place of its birth.

7

Shadows in the Particle Garden

> Particles, so tiny that one is less than insignificant, and in numbers large beyond counting, seed the garden that is our universe.
>
> —Gordon Kane[1]

Particle physics was in disarray.

"The success of quantum electrodynamics in the late 1940s had produced a boom in elementary particle theory, and then the market crashed," said Nobel Laureate Steven Weinberg[2] in a 2003 look back at the field. The abundance of new particles and forces didn't seem to fit together in any clear way. Calculations that made perfect sense for some problems made absurd predictions in others. The forces that hold the neutrons and protons together in atoms defied calculation at all. There simply weren't mathematical tools for the job.

Weinberg was one of a cast of characters who hammered physics into shape in the decades that followed the period of frustration and confusion, as Weinberg described it. The result of their efforts is the Standard Model of particle physics. It was the culmination of theoretical and experimental advances that has since been remarkably successful in accounting for all elementary-particle phenomena.

To understand how neutrinos fit in the particle garden, and some ways they don't fit so well, it takes a closer look at the Standard Model that Weinberg and other twentieth-century giants of physics built.

Particle Exchange Games

The Standard Model is a *gauge theory*. This means that the particles it describes fall into two main categories: the particles that make up the matter that we see, and *gauge bosons*, which transmit the forces that the matter particles experience.

The way the bosons transmit forces is a little like a game of dodgeball. In the schoolyard game, you can't touch opposing players with your hands. But you can transfer a lot of force by throwing a ball. If you've ever played against a much larger opponent, you may even have been knocked off your feet a time or two.

Particles play their own kind of dodgeball. Particles with electrical charge, like electrons and protons, use photons to do it instead of red rubber balls. A pair of electrons repel each other, in effect, by throwing photons back and forth.

Unlike dodgeballs, bosons can also cause particles to attract each other. It's hard to envision with the playground analogy, but imagine running a clip of a dodgeball game in reverse. In that case, the players move closer together each time a ball passes from one to the other, instead of being forced apart.

Photons don't have any charge of their own, but they are the force-carrying bosons (dodgeballs) for any particles that have electric charge (the dodgeball players). As a result, they transmit the electric and magnetic forces that are collectively known as electromagnetic forces.

There are other forces besides electromagnetism, and more force-carrying particles than just the photon.[3] The strong force that is responsible for sticking particles together in the atomic nucleus is transmitted by eight bosons known, appropriately enough, as gluons. The Standard Model also describes the weak force conveyed by three more bosons, the W-plus with electrical charge +1, the W-minus with charge −1, and the Z that has an electric charge of zero, like the photon.[4]

The weak force is particularly important for neutrinos because they are electrically neutral and don't feel the electromagnetic force. They also don't interact with gluons, and hence don't feel the strong force. But they do interact with particles that experience the weak force by throwing W and Z bosons back and forth.

The particles in the Standard Model that feel the strong force by exchanging gluons are called quarks. They have two peculiar properties: First, unlike all other particles, the magnitude of their electric charge is a fraction of the charge of an electron. Some quarks have charge that's 2/3 the size of the electron's charge; others have charge that's 1/3 the charge. Second, the quarks are bound together so tightly by the strong force that they cannot be liberated from the particles they make up. Quarks that make up protons and neutrons are forever trapped inside the particles. That's why no one has ever detected a free quark.[5] The proton has two quarks of charge 2/3, and one of charge −1/3, so its total charge is positive one. The neutron has one of charge 2/3 and two of charge −1/3, so it has a total charge of zero, as its name implies.

In addition to quarks, the Standard Model includes three particles of charge −1, called leptons, which do not feel the strong force. The lightest of these is the electron. Its heavier cousins are the muon and the tau. They all can exchange photons to participate in the electromagnetic force. They can also exchange W and Z bosons in order to take part in weak force interactions.

Each of these particles has an electrically neutral partner, the electron neutrino, the muon neutrino, and the tau neutrino. The neutrinos don't have charge, so they don't throw photons around, but they still exchange W and Z particles because they take part in the weak force.

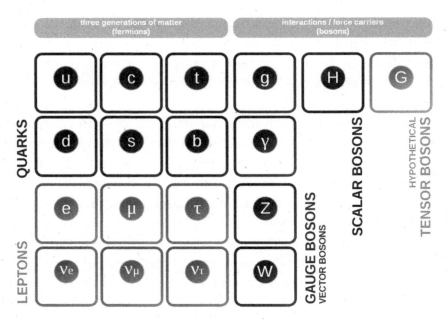

Fundamental particles of the Standard Model. The graviton, though included here, is not part of the Standard Model, but rather serves as a reminder that an ultimate "theory of everything" will have to encompass gravity as well. *Source:* "Standard Model of Elementary Particles," Wikimedia Commons (https://commons.wikimedia.org/wiki /File:Standard_Model_of_Elementary_Particles_+_Gravity.svg).

Shaping up in Spin Class

The Standard Model obeys the laws of quantum mechanics and incorporates the principles of Einstein's special theory of relativity. The particles it describes must have at least two properties: mass, which may be zero in some cases, and spin.

Spin is a form of angular momentum. In everyday life, angular momentum is always associated with rotation. If you tie a rock to a piece of rope, and swing it around your head, the rock carries angular momentum. A heavier rock has more angular momentum, as does one that is whirling faster, or one at the end of a longer rope. The direction of the angular momentum is described by the right-hand rule: Curl the fingers of your right hand in the direction of rotation; then your thumb points in the direction of the angular momentum.

Angular momentum is useful because, like energy and electric charge, it's conserved. When figure skaters start rotating, and then pull in their arms, their rotation rate increases to keep the angular momentum the same.

In the quantum world, angular momentum can be due to orbital motion comparable to the rock on a rope, but there is another source of angular momentum called spin that is intrinsic to fundamental particles like neutrinos.

These particles aren't actually spinning; they are elementary and have no internal structure, which means there is nothing to spin in the way we see things like rocks at the end of ropes or ice skaters spinning. It's just one of many examples where everyday analogies to quantum behavior can be both illuminating and misleading at the same time.

Spin, and indeed all angular momentum in a quantum theory, is quantized. That is, it can come only as an integer spin (1, 2, 3, ... and so on) or half-integer spin (1/2, 3/2, 5/2 ... and so on).[6] The fundamental unit of spin is very small. We don't ordinarily notice that spin can take only on quantum increments any more than we notice that everything around us is built of individual atoms. Quantum graininess washes out when you zoom out from the very tiny scale of atoms and molecules. It's much like looking at a beach from a boat out at sea. From a distance, it looks smooth, but is clearly built of sand grains when you're standing on it.

If you were to rotate in circles, you could theoretically measure your angular momentum in terms of the quantum units of spin, but it would be an astronomical number. It would be comparable to tallying your weight based on how many protons and neutrons are in your body. Quantization of spin and matter only becomes important when viewed up close at the level of molecules, atoms, and elementary particles.

The Standard Model particles that make up matter—the electron, muon, tau, and the quarks—all have spin 1/2. The force exchange bosons—photons, gluons, W, and Z—have spin of one.

Antimatter

The Standard Model also requires that every particle in the theory comes along with an associated antimatter version. The antimatter particles, or antiparticles, have exactly the same mass and spin as their affiliated particle, but the opposite electric charge. For example, in addition to the quarks with electrical charge 2/3 and −1/3, there are antiquarks with charges −2/3 and +1/3. These can combine to form the antiparticles of protons (antiprotons) and neutrons (antineutrons).

Likewise, the electron, muon, and tau each come with charge −1 and have antiparticle partners with charge +1. The electron's antiparticle is called the positron. There are positively charged versions of the muon and tau as well.

Some particles can be their own antiparticles. This can happen only if the particle has no electric charge. The photon and the Z boson are each their own antiparticle.

Neutrinos might be their own antiparticles, as Majorana suggested, because they have no electric charge. It's not yet clear whether Majorana was correct, and the issue is currently the subject of intensive research.

Helicity and Handedness

In addition to the particle/antiparticle distinction, there are two other properties of spin-1/2 particles that we need to consider: helicity and handedness (also called chirality). Helicity involves the spin of a particle and the direction of its motion.

In the quantum world, helicity has two possibilities because spin is quantized. A measurement of spin will result in it aligned either in the same direction of motion or in the opposite direction. If spin points in the direction of motion, the helicity is positive. That is, when you curl your fingers of your right hand around to represent the rotation, if your thumb points in the same direction as the motion, it's called positive helicity. If your thumb points opposite to the direction of motion, it's called negative helicity.

In the larger world we live in, you might expect that the helicity could have any value in a continuous range. Not so for quantum spin. It's quantized with only two possible outcomes for each measurement.

Even quantum mechanically, though, helicity is a matter of perspective. If a particle flies past you moving to the right with its spin also pointed to the right, it has positive helicity. But if you were to start moving faster than the particle, it would appear to you that the particle is moving to the left. It's comparable to overtaking a truck on a highway—the truck doesn't travel backward with respect to the road, but a child in the back seat might interpret the truck as moving backward.

If the spin of a particle is pointing to the right, it will continue to point right regardless of your relative motion. But as you overtake it, the particle will be moving to the left from your perspective. You would interpret the helicity as negative, with spin still to the right but motion to the left. That is, the helicity you see depends on the relative motion of you and the particle.

For things like photons that move at the ultimate limit, the speed of light, you could never overtake them. A photon moving past you to the right, with its spin also pointing to the right, will always have positive helicity because you can't go faster than light. The helicity of a photon can't change due to the relative motion of you and the photon, so helicity for a photon is said to be *relativistically invariant*.

Any particle that has no mass, like a photon, moves at the speed of light. It will have helicity that can't change. Helicity can be viewed as an intrinsic, fixed property of massless, speed-of-light particles. Particles with mass, like electrons and protons, always move slower than light. It's possible for helicity to look different depending on a massive particle's motion relative to you.

In the macroscopic world, handedness is related to shape—your left hand has a different shape from your right hand. That's where the origin of the term "handedness" comes from. Unlike helicity, it won't change based on relative motion. If someone drives past you on the highway and waves with their left hand, then you speed up and pass them, their left hand will still be their left hand. Everyone who sees the driver's left hand will agree that it's a left hand, no matter what our relative speed to the driver is. Handedness doesn't change (i.e., it's relativistically invariant), whether the driver can move at the speed of light or not.

Elementary particles have no structure, and hence no shape. Nevertheless, the ones that have spin 1/2 also have a quantum mechanical version of handedness. While it's analogous to the macroscopic handedness of your hands, like spin it's difficult (if not impossible) to envision for quarks and any other fundamental particles that have no structure.

For massless particles, handedness and helicity go together. A right-handed photon always has positive helicity, and a left-handed photon always has negative helicity. For particles with mass, the two properties aren't in lock step. A left-handed electron can have either positive or negative helicity.

But handedness comes with a quantum mechanical trick (a sleight-of-handedness, so to speak). Handedness of a particle can change on its own, provided the particle has mass and travels slower than light. It's still relativistically invariant because everyone sees the same handedness at any given moment. It's as if the driver on the highway switched hands, going from waving with their left hand and steering with their right to waving with their right hand and steering with their left.

In the quantum mechanical world, you never see the switch in action; you only see one hand or the other. But probability rules the quantum world: As long as you're not looking at the driver's hand, it's actually a blend of right and left. You can't say for certain which it is until you make a handedness measurement. To do that, you just have to look at it. At that point there will be a probability of you seeing one hand or the other. Wait a while between glances, though, and what starts out as a quantum driver waving with their left hand will acquire some probability of being a quantum driver waving with their right.

These properties are important, according to the Standard Model, because only left-handed particles (and right-handed antiparticles) can interact with the W and Z particles that carry the weak force. The opposite version, right-handed neutrinos (and left-handed antineutrinos) are sterile; they simply do not interact at all, except perhaps via gravity—that is, if they exist despite the lack of any clear experimental sign of them so far.

The fact that the weak interaction distinguishes between left-handed and right-handed particles came as a shock when it was discovered in the 1950s. Right-handed particles are mirror-like reflections of left-handed ones, just as your right and left hands are reflections of each other. And physicists had believed that the fundamental laws of nature must be symmetric with respect to reflection, meaning anything right-handed particles can do, left-handed ones can do as well. The weak interactions prove that this is not the case.[7]

Beta Decay and the Weak Force

When beta decay first reared its troublesome head over a century ago, no one knew the details of what was going on that would allow a proton to convert into a neutron (or the reverse). All they really could say was that the neutrino ensured that the things that had to be conserved along the way (energy, momentum, and angular momentum) were, in fact, conserved.

It's a little like knowing your Tesla runs on electricity. If that's all the information you have about your car, what happens between plugging it in at night and driving down the highway the next day is just a magical mystery going on somewhere under the sheet metal. The Standard Model lifts the hood on particles like protons and neutrons to let us see the quark motor that drives beta decay.

If you add up all the quark charges under a neutron's hood (2/3, –1/3, –1/3), you end up with a charge of zero. Neutrons have no net electrical charge, because the charges of the quarks inside them add to zero.

When the neutron undergoes beta decay, a down quark with charge –1/3 converts to an up quark with charge +2/3 by emitting a W particle. The total charge of the quarks increases by one. In the place of the neutron, with charge zero, there is a proton with charge one. But charge has to be conserved, so the W particle has a charge of –1 to balance things out. The W then decays to an electron and an electron antineutrino.

From the outside, you see a neutron turning into a proton, while an electron and neutrino fly away. When this happens in an atom, it changes to an atom of another element. That is, it turns into an atom with one more proton than the initial atom had, and one less neutron. In the case of a hydrogen atom with one proton and two neutrons, it turns into helium with two protons and one neutron. Whenever a beta decay like this happens, it moves the atom one space to the right in the periodic table of the elements. Hydrogen becomes helium, carbon becomes nitrogen, and so on.

Another variety of beta decay involves a charge –2/3 quark emitting a positively charged W particle, becoming a charge –1/3 quark, while the W converts to a positron and a left-handed neutrino. This is how a proton converts into a neutron via beta decay. The neutron is heavier than the proton, so this reaction cannot take place on its own because energy will not be conserved. However, inside certain atoms the reaction can occur due to the interactions between protons and neutrons.

The result is an atom of one element turning into an atom with one fewer proton and one more neutron. This moves the atom back to the left by one element in the periodic table, rather than to the right.

Making Mass

In the original Standard Model, neutrinos were thought to be massless. In that case, handedness doesn't change. A quantum car with no mass travels at the speed of light, and a driver who waves at you from inside a massless quantum car will never change hands. A massless, left-handed neutrino also can't change handedness. That is, a massless left-handed neutrino could never convert to a right-handed antineutrino.

All that changes if neutrinos have mass, as we now know that they do. The right-handed antineutrino produced in beta decay will evolve into a blend of right-handed and left-handed components. There are then two possibilities: The left-handed component could represent the neutrino, in which case the neutrino and antineutrino are the same particle. Or it could be a sterile antineutrino, in which case the left-handed neutrino is not related to the antineutrino by a mere change of its handedness. That would make it a distinct particle. In this book, we often use the terms "neutrino" and "antineutrino" without prejudice as to whether they really are their own antiparticles, as Majorana described, or are Dirac-type neutrinos that have separate matter and antimatter versions.

The latter possibility makes neutrinos similar to their charged electron, muon, and tau partners, which come in both left-handed and right-handed version of +1 and −1 electrical charge. Unlike the case for neutrinos, the right-handed particle and left-handed antiparticle of the electron, muon, or tau are not sterile. They do not participate in the weak interaction, but they still have electric charge, so they do interact with the photon.

The Standard Model includes an explanation of the origin of the masses of some fundamental particles. The Higgs mechanism, named after British physicist Peter Higgs, who was one of its discoverers in 1964, is an example of what physicists call spontaneous symmetry breaking.

The Beauty in Symmetry

Much of the world around us is shaped by symmetry breaking. You can see an example[8] even in the shape of a snowflake. They are born of the near perfect symmetry in clouds. If you've ever flown through a cloud in an airplane or stood in a thick fog, you may have noticed that it looks the same in every direction. That is, a cloud is a symmetric water mist. There is nothing you can see to distinguish up from down, left from right, or front from back. When a cloud cools enough to create snow, some of that symmetry goes away.

A snowflake at one place in the cloud looks different from its neighbors. If you move from one place to another, the snowflakes will be different. That is, they lack translational symmetry. In a warm cloud of mist, with no snowflakes to serve as markers, one place looks just like another. It's one reason driving in the fog can be tricky. You have a hard time telling the speed and direction you're moving in a fog bank.

Snowflakes themselves have a beautiful symmetry. But they're not completely symmetrical. They look different depending on whether you view them from the edge or on a side.

The symmetry that remains in snowflakes makes them beautiful. But it's the missing symmetry that makes them more interesting than the perfectly uniform water vapor deep inside a cloud. The drop in temperature in a cloud triggers the shift from boringly perfect symmetry to the less symmetric, but more lovely, shapes in snowflakes.

When it comes to the Standard Model, the beauty lies in the universe that existed before symmetry breaking. Unlike snowflakes, the remaining symmetry in the Standard Model contains what seem like random, and wildly disparate, masses and mixings of particles that do not appear at all beautiful now.

The higher symmetry of the early universe ensured that all the spin-1/2 particles, and all the force-carrying bosons, had zero mass. It put the weak, strong, and electromagnetic forces on an equal footing.

As the universe cooled, spontaneous symmetry breaking, due to the Higgs mechanism, kicked in. It gave mass to all the spin-1/2 particles except the neutrinos, and also gave mass to the W and Z particles that carry the weak force. They are rather heavy, the Ws being more than 80 times heavier than the proton, and the Z more than 90 times heavier.[9] It is the large masses of the W and Z particles that make their interactions short range and distinguishes the weak force.

In terms of the dodgeball analogy, it's harder to throw a heavier ball as far as a light one. For electromagnetism, photons are the force carriers. They are massless, which is as light as the particle-exchange dodgeball can be. They can be thrown to any distance, extending the electromagnetic force to infinity.

The high mass of the Z and W force carriers (very heavy dodgeballs) shortens the distance the force can act. So, the effects of the weak force are rapidly outstripped by the electromagnetic force at distances hundreds of times smaller than the size of a proton.

The Higgs mechanism acts like a kind of molasses. Particles are slowed as they move through the molasses, which is what makes them feel heavier. The amount of mass that different types of particles acquire from the molasses varies dramatically. The top quark is 350,000 times heavier than the electron, even though the Higgs mechanism is responsible for the masses of both.[10] Photons sail through the molasses completely unaffected.

The Higgs mechanism implies the existence of another force exchange boson to add to the photon, gluon, W, and Z. It's the Higgs boson. Physicist Leon Lederman famously called it the "God particle." It gives the gift of mass to matter, like Prometheus handing down fire from the heavens.

Unlike any of the other Standard Model ingredients, the Higgs has spin zero. The discovery of the Higgs was announced to great fanfare in 2012. Researchers at the CERN laboratory, which straddles the border of France and Switzerland, had found that the Higgs has a mass about 130 times that of the proton. The Large Hadron Collider (LHC), a particle accelerator 27 kilometers in circumference, is the only machine on Earth to manage to create it. With the discovery, the Higgs completed the roster of particles required by the Standard Model.

Neutrino Mass Troubles

The Higgs mechanism does not, however, naturally explain the masses of the neutrinos. It's possible to imagine an extension of the Higgs mechanism that would give the neutrinos mass, but to do that the neutrinos would have to be Dirac-type particles that have distinct matter and antimatter versions. If so, there would have to be sterile right-handed neutrinos—which have not yet been discovered—along with the left-handed antineutrinos we know about.

The only role of the sterile neutrinos would be to interact with the Higgs molasses. This would offer no understanding of why neutrino masses are so much smaller than the masses of their charged partners, the electron, muon, and tau particles.

Many physicists, instead, prefer an explanation of neutrino masses that goes beyond the Standard Model. That means building new models that will not only fill in the things the Standard Model misses. It will potentially also simplify and illuminate the Standard Model itself.

Experiments are unlikely to deliver unambiguous signposts. Rather, there are hints and clues that need to be properly interpreted. Constructing new models is a lot like cracking a code. The way forward involves sifting through a large spectrum of possibilities.

What keeps researchers on the hunt is the thrill of finally understanding what no one else has yet understood. The feeling, common across a wide range of scientific disciplines, was well described by mathematician and code-breaker Sam Blake: "There was definitely that moment, of there being this small window of time where we're the only people in the world who have seen this, and that was pretty special. And that was the pleasure of finding it out."[11]

The leading questions on the current particle physics agenda are: What type of masses do neutrinos possess? Are they Dirac particles that have distinct antimatter partners? Or are they their own antimatter partners, as Majorana suggested?

If Dirac was right, they likely get their masses from the Higgs mechanism or something similar. If Majorana was right, the mechanism by which they acquire mass must be quite different.

No one has figured out a simple way for the left-handed electron, muon, and tau neutrinos to have mass in the Standard Model. We need a somewhat more complicated strategy.

A general rule of model building in physics is that when your model fails to have the properties you want, you can fix it by paying a price in complexity: more particles, more interactions, or both.

That is the case here. One way to handle the Majorana mass problem is to propose a sterile neutrino, with a Majorana mass often designated simply as M. The sterile neutrino doesn't participate in the weak interaction. Imagining that it has mass doesn't affect anything we can measure. So far, so good.

When spontaneous symmetry breaking takes place, the Higgs mechanism can be arranged to generate a mixing between the sterile neutrino and the active one. It creates a situation that is similar to neutrino oscillation, where the flavors are superpositions of neutrinos with definite masses. The pleasant news is that if we imagine a very large value for the neutrino mass M, the other neutrino mass state must be light.

In that case, the lighter state is a superposition of active and sterile, as is the heavy state. The active and sterile states no longer have definite masses. The light one is mostly active, and the heavy one mostly sterile.

From a theoretician's point of view, it's good to have the extra neutrino be extremely heavy because that could explain why it hasn't been seen. There's simply not enough energy available in current accelerators to create it. At the same time, it is also good for the other neutrino in the model to be very light. It would be consistent with the tiny masses of the known neutrinos.

The Modeler's Privilege

The proposal of a ultra-high-mass sterile neutrino illustrates another principle of model building: From a theorist's perspective it's okay to festoon the model with complicating elements, as long as they show up at energies beyond the reach of experiment. In the case at hand, to achieve active neutrino masses that agree with the tiny masses electron, muon, and tau neutrinos seem to have, the sterile neutrino mass M must be far above the energies of any accelerator.

The extra neutrino is relegated to the high-energy Valhalla where, model builders imagine, the laws of physics may enjoy a beauty and simplicity that is not directly revealed in our earthly experiments. It is also possible that this new, heavy neutrino could play a role as a candidate for dark matter.

The situation in which one neutrino is very heavy and another very light is dubbed, aptly enough, the seesaw mechanism. It comes in three varieties, depending on exactly which new heavy particles are introduced to make it work. You can think of the Dirac neutrino with distinct matter and antimatter versions as a special case in which the seesaw is in perfect balance: The left-handed and right-handed components have exactly the same mass.

The seesaw mechanism is probably the most popular theoretical candidate for the origin of neutrino masses. But there are several others. Theoretical imagination, untrammeled by experimental constraint, can be very fertile. Ultimately, of course, the verdict rests with experiment.

The relevant experiments are extremely difficult, and it is likely to be several years before a definitive explanation for neutrino mass is at hand. When that happens, physicists will have a more powerful and complete version of the Standard Model to describe the particles and forces that make up the universe.

8
Cosmic Connections

Science progresses best when observations force us to alter our preconceptions.

—Vera Rubin, astronomer and dark matter pioneer

Particle physicists aren't the only ones with a standard model. Cosmologists focus on the portion of astronomy dedicated to understanding the origin, structure, and fate of the universe. Over the past few decades, they have constructed their own standard model to describe the universe at large. While neutrinos are an important part of the particle physics Standard Model, they aren't central to the cosmological version. Still, they almost certainly play key roles in sculpting the cosmos we see around us.

Cosmology's dominant player is gravity. The standard cosmological model is based on Einstein's theory of gravity, which he called general relativity. The central feature of general relativity is the interplay of matter and the geometry of spacetime. The presence of matter causes spacetime to curve, and the curvature of spacetime determines how matter moves under the influence of gravity.

To discuss the large-scale behavior of the universe, physicists invoke the cosmological principle. According to the principle, the universe is homogeneous (it looks the same at every point in space) and isotropic (it looks the same in every direction). Clearly, this is not true in the region of the universe around us. The solar system looks very different if you're on the Earth than it would if you were to travel to Neptune. It goes from the vacuum of space to the churning plasma in the sun and the crushing pressures of Jupiter.

Our little corner of the universe is anything but homogeneous and isotropic. Our galaxy, too, and even the local cluster of galaxies are lumpy and very different depending on the direction you look. Averaged over large enough distances, though, it all smooths out. On a cosmic level there's no special location or direction in the universe. One place is as good as another. The laws of physics are the same everywhere, and experiments don't change depending on which direction you're facing.

Another important feature of the universe is that it's expanding. In the early days of general relativity, most scientists believed the universe was static and unchanging. Einstein himself tried to construct a static model of the cosmos. This was possible only if he added a new ingredient to the theory called the *cosmological constant*, which is referred to with the Greek symbol lambda, Λ. But a decade later, Edwin Hubble demonstrated that the observable galaxies are, on average, receding from us. The universe as a whole is expanding. This led Einstein to abandon the cosmological constant. There was no longer a need to explain why the universe was unchanging, once Hubble showed that it's not.

Cosmologists increasingly began to favor the idea that the entire universe began as a single point and exploded outward. Although, at the time, there was no clear way to prove it. Astronomer Fred Hoyle was a holdout who derisively coined the term *Big Bang* as a way to dismiss the idea in lieu of an unchanging universe.

A watershed moment in cosmology occurred in 1965, when Arno Penzias and Robert Wilson were searching for the source of an annoying hum in a microwave antenna they were testing for Bell Labs in New Jersey. The interference persisted, regardless of the time of day or where they pointed the antenna. It seemed to come from every direction, all the time.

Penzias and Wilson didn't realize that 17 years earlier, Ralph Alpher and Robert Herman had predicted the radiation that the antenna was picking up. They had said that it consists of photons left over from the Big Bang. We now call the hum the cosmic microwave background (CMB). It's the oldest light in existence—and essentially provides a baby picture of the early universe.

About 400,000 years after the Big Bang, the universe had cooled from a cauldron of hot particles to the point that electrons and protons could bind to form hydrogen. Light interacts via electromagnetic forces, which means that electrically neutral hydrogen gas is transparent and that light passes through it with little effect. As a result, there were few charged particles left for photons to interact with. The Big Bang photons have been streaming through space over the billions of years since, their wavelengths stretching with the expansion of the universe.

Quantum mechanics relates the photon's wavelength to its energy: The longer the wavelength, the lower the energy and the cooler the photons. The universe has expanded by a factor of more than a thousand since the neutral atoms formed and the photons were free to move about unhindered. Today's microwave background is very low energy, corresponding to a temperature of only 2.7 degrees above absolute zero (−270.5°C).

The standard model of cosmology that describes this expanding, chilly universe with a smattering of matter in it emerged out of two separate lines of investigation. The first line is the universe we see today, as revealed by a collection of ground-based and space-based optical telescopes. Astronomers can measure the expansion of the universe by observing the velocities of distant galaxies. They can determine what the masses of galaxies and clusters of galaxies are, and how matter in general is distributed throughout the universe.

Second is the early universe. Although the microwave background is smooth, it's not perfectly smooth. Otherwise, there would be no possibility of galaxies forming as the universe evolved. The microwave background essentially carries an imprint of the early universe and yields many clues about its structure. In addition, by looking at very distant galaxies, astronomers can estimate the chemical mix in the early universe.

The combination of these various sources of information has led to the standard model of cosmology, which goes by the acronym ΛCDM or Lambda-CDM. In this case, Λ represents dark energy that drives the exponential expansion of the universe. It's a mysterious ingredient that produces similar effects to (or perhaps identical with) Einstein's cosmological constant. CDM is short for "cold dark matter," which is less exotic than dark energy, but still mysterious.

Cold dark matter is cold because the velocities of the dark matter particles are small compared to the speed of light. It is dark, which means it doesn't interact with the photons that make up light. And it is matter, because it behaves like ordinary matter when it comes to gravity, but it's distinct from any of the particles that are described in the Standard Model of particle physics.

The key evidence for dark energy was the dramatic discovery, late in the last century by two different groups who were studying distant galaxies, that the expansion of the universe is accelerating. Under the influence of ordinary matter and radiation, you might expect the expansion to be slowing down, as gravity pulls everything back together. Einstein's discarded cosmological constant could be the cause of the accelerated expansion of the universe. Or it could be something more complicated that has essentially the same consequences.

Important evidence for dark matter comes from the behavior of galaxies and clusters of galaxies. In the 1930s, astronomer Fritz

Zwicky noticed that galaxies seemed to be clustered more tightly than could be deduced from the gravity generated by the visible matter alone, and he postulated some unknown form of dark matter to make up the difference.

Another big step in the same direction took place in the early 1970s, when Vera Rubin and Kent Ford measured the way the velocity of stars swirling around in a galaxy depends on the distance from the galactic center. If the matter we can see were the whole story, the velocity should be lower for stars at the edge of a galaxy than for those closer in. Rubin and Ford found that, in fact, the velocity remains constant regardless of where the stars are in a galaxy. This suggests there's a lot more matter in galaxies than we can see. Enormous halos of invisible dark matter swaddling galaxies seem the most likely answer.

The cosmological standard model identifies the components that make up the energy of the universe and also specifies their relative proportions. Surprisingly, dark energy claims the lion's share, at roughly 68 percent, and dark matter constitutes another 27 percent. Ordinary matter and radiation, as described in the Standard Model of particle physics, contribute only the remaining 5 percent.

From our vantage point, conscious as we are only of ordinary matter, this may seem difficult to believe. Dark energy is important on enormous, cosmic scales. It speeds the expansion of the universe, but has no discernible effect at distances as small as our solar system, or even to the size of our galaxy. Dark matter does indeed pervade the galaxy, but its interaction with ordinary matter, if any, is extremely weak. Although we can ignore the mysterious dark components that make up 95 percent of the universe for almost all practical purposes on Earth, they are crucial to explaining the formation and structure of the cosmos.

What do neutrinos have to do with any of this? Assuming the as-yet-unmeasured, minute neutrino mass is roughly in the range oscillation experiments suggest, there's an intriguing numerical coincidence: There seems to be about the right quantity of neutrinos in the universe to account for the effects of dark energy. This may just be an accident of no particular significance, considering that the origin of dark energy is still a mystery.

Whether a particle of matter is hot or cold has to do with how fast it moves. Every type of particle has its own characteristic rest mass (which may be zero). The photon is massless; the neutrinos are so low in mass that it's hard to measure them; electrons are low mass, too, but much more massive than neutrinos; the proton is about 2,000 times the mass of the electron; and so on.

Einstein's equation $E = mc^2$ explains that a particle's mass is a form of energy, which physicists call the *rest mass energy*. If a particle's energy due to its motion (kinetic energy) is much greater than its rest mass energy, the particle is relativistic. That means it moves at very nearly the speed of light, and it is considered to be hot. In the opposite case, where the particle has much less kinetic energy than its rest mass energy, the particle is nonrelativistic, or cold, and moves much slower than the speed of light.

Astrophysicists run computer simulations to compare how cold or hot dark matter would have affected the evolution of the universe. The results using cold dark matter provide a good fit to the distribution of matter we actually see. Simulations using hot dark matter do not. Hot dark matter does not clump enough to create the stars and galaxies in the modern universe.

Standard Model neutrinos are dark, but their masses are tiny, so in the early universe they probably made an insignificant contribution to whatever the cold dark matter may be. Although the known, active neutrinos can't be the answer, neutrinos figure in the search for dark matter in at least three ways:

1. In addition to the active Standard Model neutrinos, theories that go beyond the Standard Model often include massive sterile neutrinos that might be an important component of dark matter.

2. Dark matter may interact with ordinary matter other than through gravity (if not, it will be virtually impossible to determine its properties). Neutrinos could be the probe that will lead to its detection, either through dark matter decaying into neutrinos or through the scattering of neutrinos off the dark matter. For example, it has been suggested that dark matter might cluster in the center of massive bodies, such as the sun. If dark matter decays, there might then be an excess of high-energy neutrinos emanating from the sun. Neutrino detectors have looked for a signal of this type, so far without success.

3. The large array of neutrino detectors that are now in operation or under construction often have capabilities that will allow them to search for dark matter directly. The detection of dark matter may occur as a byproduct of the effort to learn more about neutrinos. To date, no one has seen a definitive signal of dark matter, other than the gravitational effects it has on galaxies.

Many dark-matter candidates other than heavy neutrinos exist in the burgeoning roster of beyond-the-Standard-Model theories. From an experimental or observational point of view, there are no guideposts on where to look for dark matter. Proposals have included everything from mini black holes to new kinds of elementary particles to possible modification of the laws of gravity. A multifaceted search goes on.

Weighing Neutrinos with the Cosmic Microwave Background

Electron, muon, and tau neutrinos were too hot in the early universe to qualify as dark matter, but their tiny masses could still have left an imprint on the cosmic microwave background. The

European Space Agency's Planck satellite measured the cosmic microwave background with exquisite precision between 2009 and 2013.[1] The data provide a way to estimate the mass of neutrinos by considering how they would change the way the background looks in Planck satellite maps. In particular, if neutrino mass is too high, the cosmic microwave background would look different from the one the satellite reveals.

It's possible to estimate the maximum possible neutrino mass from images of the microwave background, but the best experiments on earth that aim to measure neutrino mass are far too crude to confirm the estimate. If the astronomical numbers are on the right track, physicists will need to improve the sensitivity of their laboratory experiments at least tenfold to have any hope of directly measuring the small neutrino masses that are compatible with maps of the cosmic microwave background.

Alternatively, there could be an error in the theories that describe the cosmic microwave background, possibly in the assumptions that go into cosmic models. For example, if you guess that the neutrino can decay, even with a very long lifetime, the upper mass limit could be considerably higher.[2] There is still much to learn as the interplay between cosmological observations and laboratory experiments continues to work itself out.

Neutrinos and the Mystery of Matter in the Universe

The combination of relativity and quantum mechanics dictates that every particle has an antiparticle. Neutrons, protons, and electrons, for example, have affiliated antineutrons, antiprotons, and antielectrons (aka positrons). Particles and antiparticles can annihilate each other, transforming their mass energy to other forms that typically end up as photons or neutrinos. If the early universe contained an equal number of particles and their antimatter partners, they all would have annihilated and none would have survived. But some of them clearly did.[3]

The fraction that survived is not large. Based on the amount of matter today relative to the number of photons, astrophysicists estimate that, on average, out of every 10 billion particles in the very early universe, there was only one more particle of matter than of antimatter. That is, almost all the matter and antimatter that existed when the universe was about a millionth of a second old has disappeared, leaving just a tiny residue that is the matter we see today.

What caused this excess of matter, however small it may have been? There are several ideas, with varying degrees of plausibility. The easiest answer is that excess matter was an anomaly of the birth of the universe: The Big Bang simply created more matter than antimatter, leaving nothing further to explain.

Cosmologists favor a scenario in which, right after the Big Bang, the universe underwent a rapid, exponential expansion dubbed *inflation*, which would have wiped out any initial imbalance between matter and antimatter, even if one existed. Immediately after inflation, matter and antimatter existed in exactly equal amounts, whether or not there had been an initial imbalance.

Another idea is that we may have been too hasty in ascribing a matter-antimatter asymmetry to the current universe. Yes, everything we see (other than occasional particles produced in cosmic rays or in the laboratory) is matter, but that's just in our own galaxy. Maybe other galaxies are made of antimatter, and when all is added up the amount of each is the same. The photons emitted by antimatter and matter are indistinguishable, so what we see in our telescopes cannot tell us whether a distant galaxy may be matter or antimatter. However, under these circumstances there should be boundary regions in the universe between large amounts of matter and antimatter, and indeed even regions where they substantially overlap. In those places, significant amounts of matter-antimatter annihilation should be taking place. This would result in abundant and distinctive types of radiation coming from the regions. But astronomers see no signs of such annihilations.

Sakharov's Solution

The preferred explanation for the asymmetry between matter and antimatter is that the matter excess developed as the universe evolved. In a classic paper in 1967, the Soviet physicist Andrei Sakharov laid down three necessary conditions for this to happen.[4]

The first condition is that some particle reactions result in a slight change in the amount of matter relative to antimatter. In most reactions, the matter and antimatter that goes into a reaction must be balanced by the amount that goes out. For every bit of antimatter created, an equal but opposite amount of matter has to be created, and vice versa. Sakharov's first condition is that this isn't always true. Sometimes there's more matter or antimatter coming out of a reaction than went in. But if the reactions in the universe that cause the excess of one are matched by reactions that create excesses of the other, there still ends up being nothing left over at all. The first condition is not enough on its own.

Sakharov's second condition is that there must not be a perfect symmetry between reactions that lead to excess matter and those that lead to excess antimatter.[5] If symmetry was perfect in particle physics, the processes that fulfill Sakharov's first condition would balance out, and there would be no residue of matter remaining to build the things we see in the universe. The laws of physics must be at least slightly asymmetrical to match up with reality.

Sakharov's third condition is that the excess matter must be created when the universe is not in thermodynamic equilibrium. If a system is in thermodynamic equilibrium, all its components are at the same temperature. The relative abundance of particles and antiparticles will remain equal as long as the system remains in equilibrium. Fortunately, the universe is expanding and cooling, which means it is in transition rather than equilibrium. As the universe developed, the reactions necessary to keep it in equilibrium slowed down until processes that lead to the dominance of matter over antimatter could become effective.

Neutrinos enter this picture in a couple of ways. If they are their own antiparticles, as suggested by Majorana, they automatically violate matter-antimatter conservation. The Standard Model includes processes that allow the asymmetry to be transferred to the other particles, like the protons and neutrons that make up atoms. This could satisfy Sakharov's first condition.

The weak interaction that the neutrino is subject to violates charge symmetry, partly satisfying Sakharov's second condition. New experiments suggest that neutrinos may have more to contribute. Researchers have now found signs that neutrinos and antineutrinos oscillate differently from each other,[6] suggesting that the particles also violate the charge-parity symmetry. The combination of symmetry violations, assuming the experimental results hold up, means that neutrinos can satisfy Sakharov's second condition.

Together with the fact that the universe is out of equilibrium, it looks like neutrinos may well be key to fulfilling Sakharov's three conditions for explaining why there's any matter at all.

A dominance of matter over antimatter in the universe that's facilitated by neutrinos is called *leptogenesis*. It's currently a favored scenario for explaining the prevalence of matter in the universe. Neutrinos are the catalysts that would allow it to happen. They are very elusive, but, in the early universe, when it really counted, it seems likely that neutrinos came through.

9

Lifting a Veil on the Universe

> Looking back on the beginnings of solar neutrino astronomy, one lesson appears clear to us: if you can measure something new with reasonable accuracy, then you have a chance to discover something important. The history of astronomy shows that very likely what you will discover is not what you were looking for. It helps to be lucky.
>
> —John N. Bahcall and Raymond Davis Jr., "The Evolution of Neutrino Astronomy"[1]

A stunning and violent astrophysical display was on its way.

After traveling through space for 160,000 years, a handful of neutrinos hailed the event on February 23, 1987, a few hours before a supernova burst into view in the Large Magellanic Cloud just outside our galaxy. It was our first glimpse of the exploding star that had briefly poured out hundreds of millions of times as much light and radiation as our sun.

The supernova was soon designated SN1987A. It's an abbreviation indicating the type of event: SN for supernova, followed by the year, and an appended letter "A" to indicate that it was the first supernova observed in 1987. Of the 18 supernovas recorded that year, it was the only one visible to the naked eye.

The few neutrinos that turned up in terrestrial detectors were a minuscule portion of the ones that came out of the 1987 supernova. A supernova goes off with the energy of a 100,000 trillion trillion fusion bombs. About 99 percent of the energy goes into neutrinos.

A supernova like SN1987A is, effectively, a neutrino bomb. The first place supernova shrapnel can turn up on Earth is in neutrino detectors.

Supernova Early Warning System

Neutrino codiscoverer Fred Reines knew that the experiments he was running in the 1960s were unlikely to pick up any sign of supernovas. Nevertheless, the meager chance of catching supernova neutrinos was enough for him and his team to make note of it, partly in jest, on an experiment located deep in an old Morton-Thiokol salt mine near Cleveland, Ohio. "In fact, we put a label on the tank, which read SNEWS,[2] an acronym for Super Nova Early Warning System," said Reines at a conference in late 1987 dedicated to SN1987A science. The experiment's 200-liter tank was filled with scintillator fluid and primarily intended to search for proton decay. The ability to detect light from neutrino interactions was a side benefit. "So we were optimistic, to put it mildly—the detector was somewhat too small for that purpose."[3]

The experiment's successor, the Irvine-Michigan-Brookhaven (IMB) detector's 7-million-liter tank of ultrapure water was installed in the Morton-Thiokol mine in 1979. There's no record of a SNEWS label being appended to the newer IMB tank. It was, however, one of the three detectors to record neutrinos from supernova SN1987A. "Immediately following [Canadian astronomer Ian] Shelton's announcement of the supernova sighting," wrote Reines, "we and the Japanese in Kamioka began a detailed search of the data. We of IMB found a burst of eight events in a 5.6 second period, a few hours (as Professor Bethe indicated it might be) prior to the visible supernova."

Eight events might seem a weak neutrino signal, but at that time IMB typically observed a couple neutrinos per day. A burst in less than six seconds was momentous. "Given a steady rate of one or two per day, the probability of getting eight events via random error in 5.6 seconds is really a reciprocal googleplex [*sic*] or so!" A googolplex[4] is a number so large that the amount of paper required to type it out would outweigh the universe by millions of trillions of times. "There's no reasonable doubt, not even an unreasonable doubt, that this might be accidental," Reines wrote.

Of the 10 billion trillion trillion trillion trillion (10 followed by 57 zeros) neutrinos produced in the supernova explosion, 25 or 30 of them were detected on Earth. In addition to the IMB neutrino signal, the Kamiokande II detector in Japan saw bursts 10 seconds apart of nine and three neutrinos. The Baksan detector in the then-USSR counted five. A detector located under Mont Blanc in France recorded five events, but these occurred three hours before the others. Either the Mont Blanc events were just an extraordinary coincidence, having nothing to do with the supernova, or else they point to something yet to be understood about either neutrinos or supernova explosions (or both).

Without a shadow of a doubt, as Reines saw it, the neutrino burst was an unmistakable starting gun for the supernova that was to follow. Although the SNEWS label attached to the prior experiment he had orchestrated in the Morton-Thiokol mine was tongue-in-cheek, it was clear with the advent of IMB and other large, sensitive detectors that an early warning system for supernovas, or any other major neutrino-producing event, was no longer a joke. While the idea of a neutrino-triggered alert was common chatter in the scientific community, it would take another 17 years for a functioning system to come to fruition.

A New SNEWS

Supernova SN1987A would have been a rare opportunity to see the death throes of a star in action from the very first moments of its demise, if only someone had been watching the neutrino detectors at the time and sent out an alert to other telescopes. A few years after supernova SN1987A, two young physicists, Boston University postdocs Kate Scholberg and Alec Habig, set about ensuring that we will never miss another nearby supernova again. They would turn to the worldwide neutrino detector community to do it.

Scholberg and Habig were not aware of Reines's acronym when the group they led settled on the name SuperNova Early Warning System, and chose the same SNEWS abbreviation that Reines thought of for their networked alerts. "It's been running since 2005 as a simple coincidence system and forwarder of information," automatically monitoring the signals from multiple neutrino observatories around the globe, explained Scholberg, who is now professor of physics at Duke University.[5]

Initially, SNEWS member observatories included Super-Kamiokande in Japan, the Large Volume Detector in Italy, and the Sudbury Neutrino Observatory in Canada. In recent years, more detectors have joined the system, including Daya Bay in China and IceCube in the Antarctic, along with new observatories alongside the Italian, Japanese, and Canadian experiments that recorded SN1987A neutrinos. Other than practice alerts, SNEWS hasn't put out a single supernova alert as of this writing. That's not surprising, considering the odds of something like supernova SN1987A happening in any given year is about one in 50.

A revised system, SNEWS 2.0, significantly expands the capabilities of SNEWS 1.0, and of course goes well beyond Reines's version, which could be considered SNEWS 0.0. There are high aspirations for the latest iteration. "It will use more information from the individual experiments to try to provide enhanced alerts to the community," said Scholberg. As additional observatories join the SNEWS 2.0 network and existing ones are upgraded, the system will be able to produce what Scholberg describes as true early warnings of supernovas, in the sense that it will send alerts before the core collapse neutrinos emerge. By looking at the timing of signals in neutrino detectors, SNEWS 2.0 can triangulate to locate the region of the sky where a supernova is about to appear. The alerts will give other telescopes time to swing around to point in the correct direction before any light arrives.

"Stars that are about to blow up," said Scholberg, "in the very last hours and days of their lives, go into new nuclear burning regimes. . . . And that produces an uptick in the neutrino production." Betelgeuse, the second brightest star in Orion, is a swollen red giant star that Scholberg and other neutrino astrophysicists suspect is ripe for going supernova. SNEWS alerts would help track the change in neutrino emissions as a star like Betelgeuse transitions to the end of its life and begins to collapse, providing a more complete picture of the mechanisms and phases of a supernova star's final moments.

Early neutrino signals will also offer insight into the end stages of failed supernovas that collapse into black holes. When that happens, a star apparently on its way to a supernova explosion will suddenly seem to wink out of existence, instead of blowing up. Currently, astronomers identify failed supernovas through broad surveys of the sky to keep an eye out for supernova candidates that show a sudden decrease in their light output. If detectors in the SNEWS 2.0 network identify pre-supernova neutrinos from a star destined to become a black hole, the signal should initially look much like a normal supernova. Nothing short of the intense gravity of a black hole can stop the escape of elusive neutrinos. That makes a sharp cutoff of the neutrino signal an indubitable hallmark of a new black hole.

Some of the observatories in the SNEWS 2.0 network are already sensitive to pre-supernova neutrinos that might come from a handful of nearby supernova candidates. Scholberg expects that the network of observatories will soon reveal pre-supernova neutrinos as far away as the center of the Milky Way galaxy.

In the more than 30 years since supernova SN1987A, the art and science of neutrino detection has progressed enormously. Even more detecting power is on the way. Should the ejecta from another supernova reach the Earth soon, perhaps from within the galaxy itself, we can anticipate many more than 25 or 30 neutrinos in detectors the next time around.

Seeing the Sky in Neutrinos

For most of human history, astronomy consisted of looking skyward to observe stars and planets with unaided eyes. The invention of the optical telescope in the early 1600s and its improvement over the next 400 years radically expanded our view and understanding of the nearby universe.

Telescopes that view the sky in wavelengths of light beyond the visible became staples of astronomy in the twentieth century. They include telescopes that scan the universe in X-rays, radio waves, infrared light, microwaves, and gamma rays. More recently, scientists have developed technologies that don't rely on light at all, including cosmic ray observatories, gravitational wave detectors, and, of course, neutrino detectors.

Each technology provides a unique picture of the universe and the objects in it. What sets neutrinos apart is our ability to use them to probe regions and objects that are fundamentally inaccessible by any other observational method. The very characteristics that make neutrinos so difficult to detect ensure that they can penetrate matter and travel cosmic distances that distort or destroy other types of signals. Neutrinos can reveal the dynamics of the sun very nearly in real time, escape the interior of supernovas to potentially provide records of some of the most violent events in the cosmos, and linger long after their creation, perhaps allowing us to someday see back to the very beginning of the universe.

We don't yet have neutrino observatories that can exploit all the information that neutrinos have to offer. But rapid advances in the past few decades, and detectors both planned and under construction, are lifting a veil on the universe that is impenetrable by any other means.

Neutrino detectors are the centerpieces of observatories that monitor neutrinos from space. Although they're among the largest experimental devices ever built, the detectors are only part of any telescope, and not the largest part by far.

Your eye may be the detector for small optical telescopes composed of lenses or curved mirrors. In radio telescopes the detector is a receiver mounted at the point where a large metal dish focuses signals from space. In either case, relatively little of the full observatory system is the detector itself. That is, for any observatory, most of the structure consists of lenses, reflectors, and filters that ensure a useful signal makes it through, whether it's destined for your eye or a neutrino detector.

For neutrino telescopes, the largest component is the Earth. Only neutrinos can survive the trip through the planet, which makes it an excellent filter to remove particles that would otherwise swamp neutrino measurements. Unlike optical, radio, and X-ray observatories that perch on mountaintops or orbit aboard satellites to look skyward, neutrino observatories are generally below ground, ice, or water. The overlaying material offers shielding from particles impinging from overhead, and the Earth below blocks interference coming from under foot. Although the detectors can reveal neutrinos raining down from space, sorting them from particles produced by cosmic rays hitting the atmosphere nearby is still tricky. Instead, many observatories primarily look downward to catch signs of the neutrinos coming up from below, after making their way through our rocky planet.

Even with the planet as a filter, most of the upward-going neutrinos don't have extraterrestrial origins. They're created primarily in cosmic ray showers on the far side of the Earth. Only about 1 percent of the neutrinos that pass through the planet originate in distant astronomical sources. Catching and studying those few that come from afar is what makes a neutrino detector a key component of an astronomical observatory.

Detectors are sometimes classified by the masses of their effective detector material. The very first experiments were on the order of tons. Borexino in Italy weighs hundreds of tons, and Super-Kamiokande in Japan is a kiloton-scale detector. The biggest neutrino detector scientists have built to date is the IceCube experiment embedded in the Antarctic ice. It consists of an array of light sensors strung like pearls on 86 cables. The cables and their detectors span a total volume under the ice of one cubic kilometer, 282 times the interior volume of the Superdome stadium in New Orleans. IceCube is the first and, so far, only gigaton-scale detector because it collects signals generated in a little more than a billion tons of ice.

To construct the IceCube array, the research team bored holes in the ice with hot water, a process that required 48 hours per hole. They lowered the detector strings down, allowing the water to refreeze and leave the array encased in ice. Although the cables extend nearly two and a half kilometers down, the detectors cover only the lower kilometer of cabling where they are spaced at 17-meter intervals. Each string includes 60 light-sensing modules, which are identified by names rather than numbers. "The modules were manufactured in the US, Germany and Sweden," says IceCube spokesperson Olga Botner,[6] "and we have modules named after islands in the Swedish archipelago, beers, birds, rivers, German counties, cars and many others. I don't recall who started [it]." Cosmic ray detectors on the ice above the array, which help to distinguish between useful neutrino signals and background noise, get their own names as well, such as Unagi, Octophobia, and Sasquatch.

When a neutrino interacts with the water molecules in IceCube's cubic kilometer of ice, the result is a daughter electron, muon, or tau particle, depending on the flavor of the incoming neutrino. Muon neutrinos are the best of the three for astronomical observations. The muons produced by energetic neutrinos travel in almost perfect alignment with the direction of the incoming neutrino, providing a distinct pointer to the source in the sky.

The light sensing modules are on the lookout for flashes that result from particles briefly exceeding the speed limit set by light in ice. Light speed in vacuum is the ultimate, unbreakable limit,[7] but in transparent materials like water, glass, and ice, light travels slower. In the ice of the Antarctic, light speed is about 24 percent slower than the speed of light in vacuum. A high-speed muon created by an incoming neutrino may briefly exceed the speed of light in the ice, but like an airplane breaking the speed of sound in air, the result is a shock wave. For a plane, the shock is a sonic boom that can rattle buildings and sometimes break windows. In IceCube, the shock that comes from a speeding particle is a cone of light named for Pavel Cherenkov, the Soviet researcher who first discovered it in 1934. The creation of light saps the energy in the speeding particles, which slows them down to the local limit while announcing the presence of the speed-breaker in the ice.

The effect happens only for charged particles like muons, electrons, taus, and their antiparticles. Neutrinos themselves don't create shock waves, regardless of their speed, because they have no electric charge.

Muons create the clearest Cherenkov signals. Electrons, as the lightest of the three potential daughter particles, scatter off ice molecules more than their heavier muon and tau cousins, which makes it difficult to deduce the directions the neutrinos were traveling as they entered the detection region. Careful analysis can narrow down electron neutrino path uncertainty to about eight degrees in IceCube. That's about the span of sky that your hand covers when you hold it at arm's length. It's an enormous and imprecise range, even as far as naked eye astronomical observations go.

Taus are 16 times heavier than muons, and more than 3,000 times the mass of electrons, but they don't live long before decaying into other particles. Unless an incoming tau neutrino is very high energy, the light cone from a daughter tau will register only in a single region of the detector array, which doesn't provide enough information to indicate direction, or even that the light burst is due to a tau at all rather than an electron.

If the energy of an incoming tau neutrino is high enough, though, it may produce a distinctive "double bang" signal, an effect predicted by John Learned and Sandip Pakvasa of the University of Hawaii.[8] The first bang comes when the neutrino interaction creates a tau and an affiliated particle cascade, and the second happens after the unstable tau travels some distance and decays, producing another cascade. Tau neutrinos with energies sufficiently high to induce discernible double bangs in IceCube are rare. A reanalysis of IceCube data from 2010 to 2017 hints at two possible tau detections,[9] but they didn't reach the high confidence levels that scientists needed to count them as true discoveries.

IceCube's primary scientific mission is to study neutrinos from distant astrophysical sources. Since the observatory's completion in 2011, it has teased out hundreds of high-energy, astrophysical neutrinos from the enormous flow of the ones created in our atmosphere. The highest-energy neutrinos the observatory has found so far are thousands of times the energy of particles produced in the world's most powerful particle accelerator labs. It's almost certain that neutrinos at those energies must come from far beyond our solar system because there's simply nothing violent enough to create them anywhere nearby.

IceCube's Neutrino Name Game

The penchant at IceCube for naming things extends beyond detector modules to notable neutrinos. The most famously named entities associated with the observatory were three ultra-high-energy neutrinos. In 2013, grad student Jakob van Santen found the numbers that the collaboration had been using to identify events unwieldy and unmemorable, so he gave the two highest-energy events nicknames based on the patterns they had produced in the detectors.[10]

Van Santen called the shorter and wider pattern Bert and the vertically elongated one Ernie, after two Sesame Street characters with distinctive head shapes. The names stuck and are occasionally used in formal scientific literature discussing the high-energy neutrinos.[11] In reality, van Santen misremembered the characters' names. Bert should have been the vertically elongated pattern, and Ernie the horizontally spread out one, instead of the reverse as van Santen named them. The collaboration continued with van Santen's theme when they designated a still higher-energy neutrino "Big Bird."

The IceCube collaborators continued naming events after Sesame Street characters for a while, but apart from the first three the names never really caught on. With subsequent names, they have strayed from the educational TV connection. Some events get names because they represent a first of their kind and are discussed a lot within the collaboration.

"It is so much easier to refer to them by their nicknames than by their real names, i.e., long strings of numbers," said Botner. "The name is usually picked by the person or the people who first see it in data."

A March 2021 discovery of a particle shower in IceCube that seemed to confirm a high-energy event known as the Glashow resonance is distinguished enough to earn the name Hydrangea. It is a landmark observation not simply because it is the highest-energy event the experiment has found so far, but also because it seems be the first indication of a long-predicted particle physics phenomenon. In 1959, Sheldon Glashow proposed that a collision between an electron antineutrino and an electron could result in the creation of a W boson, one of the particles responsible for transmitting the weak force. IceCube detectors measured the energy Hydrangea deposited in the Antarctic ice to be very close to the amount needed to make a W. Along with the pattern it left behind, it's possible that Glashow's prediction has been confirmed. More detections will be necessary to reach the level of a definitive discovery.

Detector arrays under construction in the Northern Hemisphere, with optimal sensitivity to neutrinos traveling up from the southern part of the planet, will complement IceCube's study of neutrinos coming down through the Earth from the north. Like IceCube, ANTARES (Astronomy with a Neutrino Telescope and Abyss Environmental Research project) in the Mediterranean and the Baikal Deep Underwater Neutrino Telescope in Lake Baikal in Russia monitor Cherenkov radiation from neutrino-produced electrons, muons, and taus. Instead of being embedded in ice, ANTARES and Baikal detectors are deployed on strings anchored deep underwater.

On completion of upgrades in the coming decade, ANTARES and Baikal will join IceCube as gigaton-scale systems encompassing a cubic kilometer of volume each. Placing strings underwater is potentially less challenging than boring holes in ice and may allow for repairs and maintenance that are impossible in the Antarctic system. Unlike IceCube, though, water-based arrays must contend with the light coming from bioluminescent sea dwellers and shifting conditions that might confound measurements, including temperature fluctuations, changes in salinity, and anything that might affect the transparency of seawater.

Plans are in development for a detector in the Pacific Ocean that will be triple the size of IceCube. The Pacific Ocean Neutrino Experiment (P-ONE) will potentially be located off the coast of British Columbia. If all goes well, P-ONE will be the latest element in a global network of gigaton detectors that rely on the Earth to act as a shared telescopic element to make neutrino maps of the cosmos.

Underground Observatories

Massive, gigaton detectors are key components in telescopes that add neutrino astronomy to our quiver of tools for mapping the cosmos, but smaller, underground detectors remain important for the study of neutrinos from the sun, supernovas, and other still-mysterious sources of powerful particle emissions.

Of the laboratories that found signs of neutrinos coincident with supernova 1987A, two remain active centers of neutrino research: the Kamioka Observatory in Japan and the Baksan Neutrino Observatory in Russia. The third, the Irvine-Michigan-Brookhaven detector in Ohio, operated until 1991.

Instead of grids of light-detecting modules that are features of gigaton detectors, kiloton observatories installed below ground monitor neutrino signals with light detectors that cover the interiors of the tanks filled with fluid. Super-Kamiokande relies on the same sorts of Cherenkov light that the gigaton ice and open water detectors use. It is sensitive to the higher-energy neutrinos from the sun, atmospheric cosmic ray showers, and, of course, supernovas like SN1987A that might turn up in or near our galaxy.

The Borexino detector under the Gran Sasso mountain range in Italy instead is filled with liquid scintillator fluid that creates light when jostled by radiation. Because the neutrino-induced particle showers create signals in scintillator, even for low-energy particles, Borexino can pick up neutrinos that Cherenkov light observatories can't see. That makes Borexino a leading solar neutrino observatory.

The bulk of the neutrinos coming from the sun are produced when protons fuse together to form a heavy hydrogen nucleus, called a deuteron, in the first stage of the sun's fusion process. These are known as the proton-proton process neutrinos, and they account for nine out of every 10 solar neutrinos. They are also the lowest-energy solar neutrinos. At only a few hundred KeV, they can't create daughter particles energetic enough to generate the Cherenkov light that Super-Kamiokande, IceCube, ANTARES, and Baikal measure.

Borexino is the only observatory so far to see neutrinos from the carbon-nitrogen-oxygen (CNO) reaction that facilitates hydrogen fusion to helium. The reaction dominates fusion in heavier stars, but is comparatively rare in the sun. The same is true of several other solar fusion processes. Borexino can see the neutrinos at every stage of the sun's fusion processes where they are released. Borexino's low energy sensitivity also makes it a good detector for geoneutrinos released in the decay of radioactive potassium, thorium, and uranium inside the Earth.

Radio Emission Neutrino Detectors

Some of the newest observatory concepts exploit portions of the Earth itself to effectively create detectors that dwarf even the gigaton facilities. The Giant Radio Array for Neutrino Detection (GRAND), as currently proposed, would consist of 200,000 simple radio antennas installed on mountaintops over an area of 200,000 square kilometers. The system would monitor the radio emissions that result when neutrinos smash into the atmosphere to produce particle showers. If it is ultimately built, for a relatively modest estimated cost of $226 million, GRAND will be sensitive to neutrinos a hundred times more energetic than those that IceCube can see.

The Antarctic Impulsive Transient Antenna (ANITA) currently relies on balloon-borne antennas, circling the South Pole at an altitude of 35 kilometers, to look for radio signals from neutrinos striking the Antarctic ice sheet. In effect, the entire Antarctic continent is the ANITA detector. The experiment is theoretically capable of seeing neutrinos with energies 10 times higher than GRAND. Although it hasn't definitively turned up any super-high-energy signals yet, in early 2020 ANITA was briefly the subject of a news maelstrom as a result of observations from the 2015–2016 mission. The detector appeared to show two neutrinos coming up from the ice at energies that should have been blocked by the planet below.

An exaggerated news story reporting that the signals might be interpreted as signs of a parallel universe briefly made ANITA a worldwide tabloid sensation.[12] There's no clear explanation for the anomalous signals as yet, although there's the possibility that ANITA's successor, the Payload for Ultrahigh Energy Observations (PUEO), will clear up things thanks to an increased number of antennas and improved sensitivity. It probably has nothing to do with parallel universes, but we should know one way or the other soon.

Studying the Sun

Modern humans were just emerging 170,000 years ago, making tools, inventing clothing, and learning to fish. Meanwhile, the sunlight that shines on you today was being born in the depths of the star at the center of our solar system.

By human time scales, the photons that come to us from the sun now are the relics of ancient fusion reactions in the solar depths.

The energy that makes the sun shine is produced in the dense core that extends to about a quarter of the sun's radius. Fusion reactions release gamma rays that diffuse out from the core and wend their way through the sun's plasma of electrons and protons.[13] Along the journey from the core to the surface, each gamma ray is converted into millions of lower-energy photons. When they finally reach the outer layer of the sun, known as the photosphere, light escapes into space, giving us the familiar, shining star that the Earth orbits.

The neutrinos that stream out from the sun, on the other hand, start with the very same fusion reactions that ultimately produce light, but escape from the core almost entirely unhindered. They arrive at the Earth eight and a half minutes after they're created, which means the neutrinos provide a snapshot of the sun essentially as it exists now, in contrast to the photons that paint a picture of solar fusion eons ago.

The neutrinos that Ray Davis measured are born in the stages of fusion in the sun that results when an isotope of boron decays, releasing a positron and a neutrino. They're comparatively energetic neutrinos, which is what made them observable in the detector Davis built in the Homestake mine in South Dakota, but they account for only one in 10,000 of the sun's neutrino output.

Several detectors around the world are sensitive to the proton fusion neutrinos, but Borexino was the first experiment capable of identifying them specifically. Comparing the solar neutrinos arriving today with the photons emerging from the sun allows us to understand how the fusion engine that drives the sun has changed over the millennia. The answer, revealed by a Borexino paper published in 2014, is not much.[14]

Modern neutrinos and ancient photons together show that the sun has been remarkably stable for over 100,000 years. Neutrinos coming from later fusion stages in the sun have confirmed models of solar fusion and provided new insights into the materials that comprise the sun's core.[15]

Detecting solar neutrinos directly can give us only a snapshot of the sun in its current state. All the neutrinos that flowed from the sun over the last 4.6 billion years since the sun formed are lost to us. There may be, however, traces that the particles left behind as they interacted with materials in the Earth. In 1990, University of Washington physicist Wick Haxton proposed that changes in the sun's luminosity over time would result in corresponding changes in neutrino flow.[16] Most of the neutrinos would, of course, pass through the planet with no effect, but some would have interacted with xenon trapped in rocks and created an exceedingly rare form of the element. By measuring the fraction of xenon that consists of the rare variety, and using geological techniques to determine the ages of the ores where the isotopes occur, it would be possible to obtain a record of the sun's output over billions of years.

Unfortunately, Haxton noted, enormous quantities of mineral ore would need to be examined, under conditions where they are painstakingly protected from contamination. The neutrino-modified xenon would likely make up less than a tenth of a percent of an element that's extremely rare to begin with. It's an experimental challenge that Haxton felt was prohibitive in 1990, and it probably remains that way today. Still, there's at least the potential that we may someday read the record of solar activity, spanning nearly the entire lifetime of the sun, that neutrinos have etched in the rock of our planet.

Messengers from beyond the Galaxy

The highest-energy neutrinos that turn up in experiments like IceCube in the Antarctic and ANTARES in the Mediterranean Sea are thought to come from outside of our galaxy. The sources are known generally as cosmic accelerators because they must produce extremely high-energy neutrinos in much the same way that research accelerators do on Earth. The specific origins of the energetic particles are not entirely clear, but astrophysicists guess they're produced in pulsars, gamma ray bursts, the remains of supernovas, or matter falling into black holes.

In particular, supermassive black holes that are millions to billions of times the mass of our sun appear to be the drivers at the cores of galaxies, which are known as blazars. They're one type in a class of galaxies that are called *active* because much of the light they emit is not due to stars. Instead, it comes about when matter spirals in toward the black hole at a galaxy's center, spewing out energy along the way in the form of light and radiation.

Blazars, in particular, are active galaxies that emit jets of ionized particles and light toward the Earth, like cosmic lighthouses sweeping their beams past us. (There are certainly many with jets that point in other directions, but we don't see them, and can't identify them as blazars.) As we see them from Earth, blazars fluctuate dramatically in brightness and intensity.

Highly energetic neutrinos are among particles we expect to be spewed out from blazars' cosmic accelerators. In 2017, that expectation appears to have been confirmed when a neutrino of 290 trillion electron volts turned up in the IceCube detector.[17] That's twenty times more energetic than the particles in the Large Hadron Collider, which is the highest energy accelerator on Earth. The path of the neutrino through the detector appeared to indicate that it originated at a blazar 5.9 billion light years away. It would be the first time that scientists have identified a particle coming from a specific source far from our galaxy. Up to that point, all previous neutrinos with known origins came from the sun, supernova SN1987A in the Large Magellanic Cloud, or natural and artificial sources on the Earth. A review of prior IceCube neutrinos found that particles from the blazar had been turning up in earlier years, although the connection was not made previously because the galaxy hadn't been flaring at those times.

Big Bang Relics

When you gaze up in the sky, you're looking back in time. Because its speed is not infinite, light takes time to travel from distant objects to our eyes. If you look at Jupiter through a telescope, you will see its colorful, turbulent atmosphere as it was between 32 and 54 minutes ago, depending on where the planets are in their respective orbits. The light from the nearest stars to us shows the Alpha Centauri triplets in their arrangement four years and four months ago. And when you see the Andromeda galaxy, the most distant feature in the sky that's visible with the naked eye, you're observing it when it was 2.5 million years younger than it is today.

In principle, if you had a sensitive enough observatory, you could see back to the very beginning of the universe itself—just not with light. Until about 400,000 years after the Big Bang, the matter in the universe was very hot and dense. Electrically charged particles, mainly protons and electrons, acted like a giant, opaque fog that trapped the light within it. As the universe expanded and cooled it became mostly transparent, which allowed light from that time onward to travel through space to our telescopes.

We can no more peer inside the hot and dense earlier stage of the universe, regardless of how sensitive and sophisticated our photon-based telescopes might be, than you can look inside a looming storm cloud with your naked eye. The farthest back in time an observatory can see with photons, whether they are visible light, infrared, radio waves, X-rays, or gamma rays, is to the stage of the early universe known as the surface of last scattering.

There are two ways to see through to the other side of the surface of last scattering, and potentially to the very beginning of the Big Bang birth of the universe. The first of these is gravitational waves. Gravitational wave antennas like the Laser Interferometer Gravitational Wave Observatory (LIGO) and more sensitive detectors on the horizon may be able, in coming years, to discern the hum of primordial gravitational waves from the Big Bang amid the din of exploding supernovas and crashing black holes and neutron stars.

The second potential source of information is the cosmic neutrino background. Like its more famous cousin, the cosmic microwave background, the cosmic neutrino background is a relic of the early universe. Although similar in some ways, there are several important differences. The cosmic microwave background is composed of photons. The cosmic neutrino background is, as its name implies, composed of neutrinos.

The microwave background is relatively easy to detect; Penzias and Wilson discovered it accidentally while trying to track interference in their microwave antenna. The neutrino background will be extraordinarily difficult to detect, because the low-energy neutrinos of which it is composed are so weakly interacting.

The microwave background we see was formed about 400,000 years after the Big Bang. The neutrino background harkens back to about one second after the Big Bang. If the neutrino background could be studied, it would reveal secrets of the early universe that cannot be learned in any other way, except perhaps from the spectrum of primordial gravitational waves.

The microwave background today is cold, about 2.7 degrees above absolute zero, the result of the expansion of the universe, which stretched the wavelength of the photons. According to quantum mechanics, the longer the wavelength, the lower the energy, a property universally shared by all matter and radiation. The neutrino background is thought to be even colder, at 1.95 degrees above absolute zero. This is because neutrinos stopped interacting with the rest of the matter in the universe, as electrons and positrons continued annihilating into photons. The annihilations heated up the cosmic microwave background, but not the neutrino background.

The density of the microwave background is a little more than 400 photons per cubic centimeter. It's likely that there are just shy of 350 neutrinos and antineutrinos per cubic centimeter in the neutrino background, assuming they are equally distributed in flavor and between particles and antiparticles. If there's significant neutrino-antineutrino imbalance, the density might be higher.

One proposed experiment, the Princeton Tritium Observatory for Light, Early-Universe Massive-Neutrino Yield (PTOLEMY), could reveal the ubiquitous primordial neutrinos by looking for electrons with slightly elevated energy coming from a sample of tritium. Usually, when tritium decays naturally, it converts to helium and emits an electron and an antineutrino. Because there are three end products, the energy of the decay is shared three ways. The electron ends up with a spectrum of energy, instead of one specific value. This is the very sort of reaction that provided indirect, but compelling, evidence of neutrinos in the first place.

If, on the other hand, a tritium atom absorbs a primordial neutrino, it will decay to a helium atom and emit a lone electron with a fixed and specific energy. That energy is higher than the peak of the electron energy that comes from natural tritium decay.

The energy accounted for by the neutrino in beta decay gets added to the electron energy instead, when there's no neutrino to carry it away. As a result, PTOLEMY offers the added benefit of measuring neutrino mass by checking to see how much extra energy the electron has in a decay induced by a primordial neutrino.

In order to provide a large enough signal, the PTOLEMY researchers hope to use a 100-gram sample of tritium. That's a quarter of the annual worldwide of supply of commercially available tritium, costing roughly $30,000 per gram.

Once the tritium is assembled, distinguishing between the electrons coming from natural tritium decay and the ones induced by relic neutrinos will be a challenge. The energy associated with the electron neutrinos mass is small, requiring a precision of one part in 50,000 in the energy measurement to see it.

An approach that could overcome the difficulties that slow, cold, relic neutrinos present is to accelerate tritium atoms to high velocities. For high-speed atoms, the otherwise lethargic Big Bang neutrinos would appear to be at higher energy, relatively speaking. It's a little like flying a kite on a windless day—if the air isn't moving, you can still get a kite aloft by running and pulling the kite along behind you. In essence, an experiment similar to PTOLEMY, except performed with tritium atoms careening through an accelerator, would increase the probability of scattering by moving at high speed through the cosmic neutrino background. It would make an extremely high-speed neutrino wind, as far as the tritium atoms are concerned. The experiment would take accelerators that reach hundreds of times the energy of any accelerator built to date.[18] So, it's unlikely to happen any time soon.

Another strategy for detecting the neutrino background relies on one of the cornerstones of quantum mechanics: wave-particle duality. At very small scales, where tiny things like atoms and neutrinos reside, the distinction between particles and waves is not clear cut. A neutrino, photon, neutron, or any other tiny object can sometimes act like a discrete particle, comparable to a bullet, and sometimes like a wave spread out on the water. The wavelength, in turn, is short for particles with lots of energy. Low-energy particles have long wavelengths associated with them.

Big Bang relic neutrinos are so cold that their wavelengths are about a millimeter. This is important because particles can scatter from objects comparable in size to their wavelength. An effective target for a relic neutrino should be about as big as a mustard seed. The probability of a neutrino bouncing off a target goes up with the square of the number of subatomic particles in it. Considering that a cubic millimeter of matter has trillions of trillions of atoms in it, the odds that a cold relic neutrino will ricochet off it are much higher.

If the target is sitting still relative to the neutrino background, neutrinos will bounce off all sides of it equally, and the net effect would be zero. It's the same reason you don't feel the atmosphere pushing on you: The pressure is the same on all sides, and the total forces cancel out.

But we're not sitting still in the neutrino background. We're riding on the Earth, which plows through the neutrino background at about a billion kilometers per hour. Still, the forces on anything moving through the neutrino background at those speeds are far below the sensitivity of any existing technology. To date, no credible experiment to measure Big Bang neutrinos scattering from millimeter-sized objects has been proposed. "We therefore have the tantalizing and frustrating situation that the big-bang model predicts that the Universe is filled with the essentially undisturbed neutrino remnants of the very early universe," wrote Paul Langacker, Jacques Leveille, and Jon Sheiman in a paper on detecting the ancient particles.[19] "The neutrino sea may profoundly affect the structure and formation of galactic clusters . . . yet these relic neutrinos are essentially impossible to detect by any conventional means. Clearly, any encouraging new approach would be very exciting."

That was in 1983. We're still waiting, as each of us spends our lives filled with tens of millions of the oldest particles in the universe.

From Detector to Observatory

Astronomical observatories are often associated with telescopes that image discrete galaxies, stars, and planets. Identifying neutrinos of astronomical origins with specific sources, however, is difficult. Early detectors were poor at narrowing down the direction neutrinos travel, and instead relied on timing information in conjunction with other observations to deduce neutrino sources.

In the case of supernova SN1987A, the bursts of neutrinos that turned up in multiple detectors around the world were associated

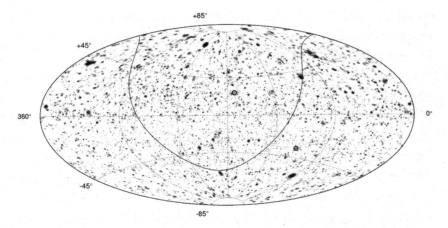

A map of the sky as seen in neutrinos. Neutrinos come from all directions, and their origins are usually unknown. The circles identify hot spots that seem to indicate specific neutrino sources. *Source:* Image courtesy of the IceCube Collaboration.

with the stellar explosion because of their timing. None of the detectors at the time was capable of identifying the neutrino flight direction with precision sufficient to serve as a conventional observatory. But the likelihood that the uptick in detections that appeared in major neutrino detectors on February 24, 1987, was the result of random fluctuations was vanishingly small. The coincidence in the timing of telescopic and neutrino observations was the only way to make the association between SN1897A and the signals in neutrino detectors.

That all changed when IceCube captured the blazar neutrino from the Orion constellation. "When I go out at night and look at the sky, I make a map using light beams I detect with my eye," says IceCube principal investigator Francis Halzen.[20] "So, what we do is exactly the same thing, but we detect neutrino beams." Halzen describes early neutrino maps of the sky as uniform and featureless with some emerging, but statistically insignificant, signs of individual neutrino emitters. "Then this one source started to appear which was this supermassive black hole, which we subsequently discovered in this multi-messenger campaign, using other telescopes." As Halzen sees it, IceCube's discovery is the first confirmed neutrino star other than the sun.[21] "The development is very similar to gamma ray astronomy from the ground. For years, they had only one source, which was the Crab Nebula. They optimized their instruments and built more, and now they're doing astronomy. So, I hope the same will happen with neutrinos," said Halzen. "In fact, I have no doubt."

10

Made-to-Order Neutrinos

Our need will be the real creator.

—Plato[1]

The accelerator at Brookhaven National Laboratory was too good. In the 1960s, the Alternating Gradient Synchrotron (AGS) was the leading machine for supplying high-energy protons for experiments. It could produce beams of particles reaching 33 billion electron volts, more than 10 percent higher than the second-best accelerator, the Proton Synchrotron in Europe, and five times the energy of the previous generation of world-class accelerators.

It was also far more energy than Leon Lederman, Melvin Schwartz, and Jack Steinberger could tolerate if they were to prove that neutrinos come in more than one flavor. The three physicists planned to slam protons into a target made of beryllium to create a beam of neutrinos. The collisions would produce showers of other particles as well, including muons, which threatened to swamp the signals they were after. Their multipronged solution involved shielding, precise timing, and intricate analysis.

But first, they needed to turn the energy down. They would hunt the muon neutrino by running the world's premier particle accelerator at less than half its peak energy.

Forging Neutrino Beams

Naturally occurring neutrinos are just one source of the particles that physicists study. We get them from radioactive elements, the sun, or the cosmos. Another source is fission in nuclear reactors, which produce electron antineutrinos in profusion. Beams of neutrinos generated in particle accelerators are a third source.

Physicists take what they can get when it comes to naturally occurring neutrinos. The sun and stars burn on their own schedules. Nearby supernovas appear both randomly and rarely. Many cosmic sources remain mysterious and unpredictable. While we can accumulate radioactive elements in bulk to study the neutrinos they emit, there aren't easy ways to tweak the properties of the emerging neutrinos.

Reactors offer a bit more control. Their primary purpose is to generate energy and to produce weapons-grade materials. Neutrinos are typically a byproduct instead of an end goal of reactors. As a result, nuclear reactor operations are rarely in neutrino researchers' hands.

Last, physicists use accelerators to make beams of neutrinos with properties they can exercise control over. Neutrino beams are research tools tailored to explore the frontiers of physics without being subject to the whims of nature, power companies, or nuclear weapons manufacturers.

Particle beams of any kind begin with electric fields to accelerate charged particles. They are typically protons and antiprotons, or electrons and positrons, or electrically charged atoms. Neutrinos have no electric charge, which means they can't be accelerated with electric fields. Instead, they must be produced indirectly.

Lederman, Schwartz, and Steinberger pioneered neutrino beam experiments in their 1961 efforts with Brookhaven's AGS accelerator.[2] Their strategy for making a neutrino beam, used to this day with various improvements, was to direct the high-energy beam of protons from the accelerator onto a target to make fast-moving pions. A pion is an unstable particle made of a quark and an anti-quark, and the experiment included a 20-meter path behind the target where many of the pions decayed.

Pions come in three types: positively charged, negatively charged, and neutral versions. The positive pions decayed to positively charged muons and their accompanying neutrinos; the negatively charged ones produced negatively charged muons and their accompanying antineutrinos. (Neutral pions typically decay into two photons, and didn't contribute to the neutrino study.)

The mélange of particles then encountered a thick block of shielding, composed of steel that had been rescued from a decommissioned, pre–atomic age battleship to ensure it would be free of trace radioactive contaminants. The steel filtered out other particles, including muons and any pions that hadn't decayed yet, leaving primarily neutrinos and antineutrinos in the beam. A neutrino detector behind the shielding had to be sufficiently massive to catch at least some of the neutrinos. It also had to be able to discriminate between the electron and muon flavors. The researchers started with about a 100 trillion or so neutrinos and antineutrinos from the accelerator. Just over 50 turned up in the detector.

When the neutrinos interacted inside the detector, they produced charged particles. The question was: Were these particles only muons or were they a collection of muons and electrons (or positrons)? If only muons remained, then the neutrinos that produced them could be labeled as muon neutrinos, distinct from the electron variety. If it was both, then the neutrino would have to be a single species, indifferent to flavor. The experiment conclusively showed that the neutrinos produced in pion decay are muon neutrinos, unlike the reactor variety that Cowan and Reines had detected six years before. It was a groundbreaking, and Nobel Prize–winning, experiment proving that neutrinos came in at least two distinct flavors.

We now know that neutrino flavors can indeed transform into one another. Had the experiment taken place over a longer distance or at higher energies, Lederman, Schwartz, and Steinberger would not have found the pure muon neutrino beam that led to their Nobel Prize. But, given the energies in the Brookhaven experiment, the relevant oscillation lengths were anywhere from tens to hundreds of kilometers, which meant muon neutrinos were unlikely to register as electron neutrinos so shortly after their creation.

A similar experimental effort set out reveal the existence of the tau neutrino three decades later. It would require a particle beam twenty-five times more energetic than the one in the Brookhaven machine, and a much more complex detector, shielding, and magnet arrangement. The DONUT (Direct Observation of the Nu Tau) detector ultimately uncovered only four tau neutrino candidates from the beam at Fermilab's Tevatron accelerator after months of data collection and years of analysis. By the summer of 2000, it was definitive enough to add the tau flavor to the neutrino lineup in the Standard Model.

So far, the tau neutrino discovery is the only one of the three neutrino flavors that hasn't resulted in a Nobel. Martin Perl had previously led the effort to find the tau particle that's the heaviest sibling in the Standard Model triplet of leptons that includes the electron and muon. Perl's tau particle discovery would earn him a half share in the Nobel Prize in 1995, with the other half going to Reines for the initial electron neutrino discovery. The existence of the tau particle made the discovery of the tau neutrino seem destined to happen eventually. After all, if the electron and muon came with their own affiliated neutrinos, then the tau should too. Confirmation that the tau neutrino exists never generated the sort of excitement that followed the electron neutrino and muon neutrino discoveries.

Accelerators continue to play leading roles in neutrino physics. In 2020, the Tokai to Kamioka (T2K) accelerator experiment in Japan appears to have shown that neutrinos and antineutrinos oscillate differently from each other. If confirmed, it will help solve the mystery of the origin of matter in the universe.

T2K is a long-baseline-distance experiment. An accelerator in Tokai generates a neutrino beam directed toward the Super-Kamiokande detector in the Hida Highlands 295 kilometers to the west. T2K researchers can select beams that are predominantly either muon neutrino or muon antineutrino by using magnetic fields to deflect the charged particles that are the initial source of the neutrino beams. When they choose positively charged pions, they end up with beams of muon neutrinos. If they magnetically select negatively charged pions, they get beams of antineutrinos. They then compare the rate of muon neutrinos oscillating into electron neutrinos to that of muon antineutrinos oscillating into electron antineutrinos. Differences between the two oscillations imply that neutrinos may have catalyzed the dominance of matter over antimatter, which makes our existence possible.[3]

The discovery is not yet conclusive. The NOvA (NuMI Off-Axis v_e Appearance) experiment at Fermilab has similar capabilities to T2K, but the data from NOvA so far are at odds with the T2K results.[4] The data to settle the disagreement may ultimately come from Fermilab.

The Fermilab neutrino program involves both short and long baseline elements and employs two different beam lines. The short baseline cluster, sited in and around Fermilab itself, makes use of the Booster Neutrino Beam (BNB) and includes MiniBooNE and its sibling MicroBooNE, both designed to probe the anomalies first suggested by LSND at Los Alamos, and possibly to discover evidence for a fourth, sterile, neutrino.

Fermilab also has a long-baseline program in operation, using the Neutrinos at the Main Injector beam (NuMI). The far detectors for the Minos and Minos+ experiments were located in the Soudan mine in Minnesota, a distance of 735 kilometers away. A successor experiment, NOvA, is 810 kilometers away from Fermilab, which is slightly off-axis from the NuMI beam, giving it a different energy range from the others.

All these experiments are designed to study parameters that determine neutrino oscillations. The Minerva experiment at Fermilab used the NuMI beam to study neutrino interactions with nuclei.[5]

In addition to the Minnesota experiments, Fermilab's flagship long-baseline experiment, the Deep Underground Neutrino Experiment (DUNE), is under construction at Fermilab and in the Homestake Mine in South Dakota, now known as the Sanford Underground Research Facility (SURF), the same location where Ray Davis built his solar neutrino experiment many decades ago.

The near detector at Fermilab will have three components, one of which will sit directly in the neutrino beam to continuously monitor its properties. The other two will be mounted on tracks, giving them the capability to move across the beam, allowing them to sample a range of beam energies. At the near detector, the neutrino beam will be so intense that even with the weak neutrino interactions there will be a high event rate that the detectors will have to disentangle. The near detector itself will have the ability to search for sterile neutrinos and other neutrino interactions that aren't part of the Standard Model.

The far detector will consist of four modules located 1,500 meters underground, containing a total of almost 70,000 tons of liquid argon as the medium in which the neutrinos will interact. A new cavern is being dug to contain them, involving the excavation of about 800,000 tons of rock. The detectors will be fed by an upgraded version of NuMI, located at Fermilab 1,300 kilometers away. Two detector prototypes have been constructed, one involving only liquid argon and another that operates with regions of both liquid and gas, and are under study at CERN.

DUNE will be the largest science experiment ever built in the United States, assuming it can run the funding gauntlet.[6] Even

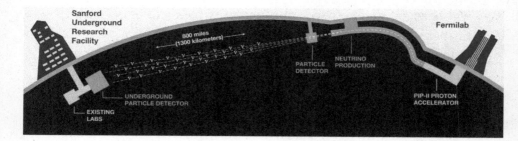

The flying ν symbols indicate neutrinos traveling through the Earth from Fermilab in Illinois on the right to the Sanford Underground Research Facility in South Dakota on the left. *Source:* Fermilab.

at the early construction phase it's under the auspices of a global collaboration of over 1,300 scientists. The immense effort means that it will require international support, much like the $20 billion International Thermonuclear Experimental Reactor that's being built to develop fusion reactor technology in France, or the equally costly Large Hadron Collider near Geneva where the Higgs particle was discovered.

If all goes as planned, DUNE should be recording data sometime in the middle of the 2020s. It will eventually help round out the complete set of neutrino parameters and look more closely at the signs of matter-antimatter asymmetry turning up in T2K and NOvA. While primarily a neutrino experiment, DUNE will also be able to contribute to searches for dark matter and proton decay.

Neutrino Factories

Next-generation neutrino beams may turn to muon accelerators to create intense and high-energy neutrino factories.[7] Protons will still be needed to create pions as in the early Brookhaven neutrino beam experiment. Instead of letting the particles coast and eventually decay to create a beam of neutrino and antineutrinos, the pions will be given time to decay into muons, which will be accelerated further.

In one of the most discussed schemes, the high-energy muons travel on to a racetrack-shaped storage ring, to be guided and focused with magnetic and electric fields. As the muons break down in the straight sections of the racetrack, the neutrinos will escape because they aren't captive to the fields that keep the muons inside the ring. The result will be powerful beams of neutrinos and antineutrinos.

An important difference between high-energy proton machines and ones that use muons is the relative simplicity of the particles. Protons are made of quarks. Muons are fundamental particles that don't have any internal pieces. Proton experiments are messy, and the energy gets spread among the interior parts. Muon experiments would be cleaner because there's just one piece. Smashing protons together is like throwing bags of marbles at each other. Muon colliders are more like bouncing individual marbles off one another. The muon interactions would be both easier to understand and make better use of their energy[8] than protons.

The neutrinos that result from the muon decays would be, as Patrick Huber of Virginia Tech describes them, "the mother of all neutrino beams."[9] These high-energy neutrino factories would give us the ultimate look at oscillations. They could resolve differences between existing experiments and provide the clearest signs of sterile neutrinos yet.

The beams would be so extreme that they also present a novel radiation hazard.[10] The greatest risk is not due to the neutrino beams themselves, but instead comes from the secondary radiation that results when neutrinos interact with atoms and create showers of dangerous ionizing particles. When the beams strike an obstacle, like the ground or shielding around an accelerator, they produce showers of other particles thousands of times more hazardous than the neutrinos.

One way to reduce the hazard of a neutrino beam is to dispense with shielding entirely. Instead, it would be safer to give the beam an unobstructed escape route, with nothing in the way to release hazardous secondary radiation.[11] If you were to wander into the path of a pure neutrino beam, interactions with your atoms would lead to a burst of particles out of your body, but relatively little of the radiation would end up in you.[12] You'd be better off being hit by a neutrino beam directly than being struck while hiding behind even a mountain of lead. (It might not be wise to be in line behind someone standing in front of a pure neutrino beam, though, where you would face the secondary radiation coming out of them.) It's not that high-energy neutrino beams are inherently safe, just that the shielding that protects people from other sorts of radiation coming from accelerators makes neutrino beams much more dangerous.

Serving up Reactor Neutrinos

Reactors provide less flexibility than accelerators, but their benefits have been undeniable since the very first successful neutrino experiment when reactors saved Cowan and Reines from the trouble of dropping El Monstro down a shaft near a nuclear bomb test.

One of the limitations in reactors is selection of neutrinos they produce. Inside a nuclear fission reactor, uranium and plutonium atoms break apart into lighter atoms, releasing energy in the process. The fragments are usually unstable and break down further, often via beta decay, which leads to the release of an electron and electron antineutrino. As this cascade proceeds, each original fission reaction produces several antineutrinos, typically half a dozen or so, making nuclear reactors sources of ample antineutrinos. Unlike neutrinos from accelerators, reactor neutrinos can't be easily tuned in energy and intensity.

Reactors are expensive, and usually are built with purposes other than neutrino research in mind. The primary applications are for energy generation or nuclear weapons production. That leaves researchers little say in operating reactors. Cowan and Reines managed to arrange for the reactors in their experiments to be powered on and off on request to check to see if their measurements were truly revealing neutrinos from nuclear fission. But it's not the sort of control that physicists typically have over commercial power plants or nuclear weapons programs.

A new generation of reactor experiments has been active over the past decade or so. Among these are Daya Bay in China, RENO in South Korea, and Double Chooz in France. All three have antineutrino detectors situated at two different distances from their respective reactor complexes. By comparing the rates of detection in the near and far detectors, physicists can tell how the antineutrinos are oscillating.

If sterile neutrinos exist, reactors would be a likely place to find them. The oscillation of the active neutrinos produced in the reactor into undetectable types of sterile neutrinos would show up as a shortfall in the flux coming from the reactor. The short baseline distances of most reactor experiments would be ideal for searches for light sterile neutrinos.

For several years, some reactor experiments showed just such a deficit, averaging about 6 percent. The situation was reminiscent of the solar neutrino problem, where the question was whether the solar models or the neutrinos were at fault. Here the issue was whether the reactor neutrino calculations were right. Recent analyses offer a better understanding of reactors. As it stands, no sign of sterile neutrinos remains in reactor studies.[13] The limits are stringent enough that they are at odds with gallium experiments that still show hints of sterile neutrinos. It's a dramatic tension between two different ways of tackling the same question. Which is right is still to be seen.

Homecooked Neutrinos

With neutrino detectors operating at several reactors around the world, and major accelerator programs in planning or construction phases, the prospects for new discoveries in the coming decade are bright.

There's always the hope that a nearby supernova will explode into sight, or some mysterious cosmic source will send more ultra-high-energy neutrinos our way. In the meantime, experimenters have plenty to learn from homemade neutrino sources.

11

A Coherent Tale

There is no doubt that scientific advances depend not only on new ideas, conceptual leaps and paradigm shifts, but also to a large extent on technological advances that make these steps possible.

—Editorial, *Nature Cell Biology* 2 (2000): E37

The U.S. Navy had a neutrino problem. In the mid-1980s, it was a mortal threat to the fleet of submarines that were prowling the oceans in near total secrecy.

It had been enough of a concern years earlier that U.S. intelligence had compiled a top-secret report titled "Soviet Antisubmarine Warfare: Current Capabilities and Priorities." The 1972 analysis had concluded that the Soviets were unlikely to build a sub-tracking neutrino detector.[1] Despite the fact that neutrinos from nuclear submarines could not be shielded or suppressed, the report noted, a detector that could pick up neutrinos from submarines at any significant distance would have to be enormous. Even if such a detector was built with the most sophisticated technology, the information it could provide would be of dubious value; although it might be able to tell a nuclear submarine existed, it couldn't determine the sub's location or the direction it was moving.

The report's reassuring conclusions appeared to be shattered in 1984 when Joseph Weber, a physics professor at the University of Maryland, published a paper in the journals of the American Physical Society.[2] He claimed to have created a neutrino detector 10 million trillion times more sensitive than any built before. What's more, it could measure the momentum of the neutrinos, which meant it could tell the direction they were coming from, potentially allowing them to pinpoint the location of a nuclear-powered sub anywhere on the planet.

Previously, Weber had primarily been known for his attempts to measure gravitational waves—an effort he claimed was successful, but which evaded duplication by other researchers. Weber's spouse, astronomer Virginia Trimble (University of California, Irvine, and the University of Maryland) remembers that the late physicist's change of focus from gravitation to neutrinos came about as they were sitting down to lunch with Richard Feynman. "Some time in [the] 1970s, he and I were up at Caltech," wrote Trimble in a 2021 email. "Joe was attempting to explain the evidence that the [gravitational] signals being recorded by his bar detectors were real. Feynman with his usual well-known (for its absence) patience said something like 'enough with the gravity waves already. Go look for neutrinos or something.' Joe took this as serious advice (well perhaps it was) and settled in to calculate how he might look for reactor and solar neutrinos with a much larger cross section." Trimble recalls that the Defense Advanced Research Projects Agency (DARPA), the agency responsible for developing emerging military technologies, "supported the program for some time because of the potential for detecting nuclear submarines very quietly from a distance."

Weber's experiment involved a perfect sapphire crystal and an exquisitely sensitive measurement device called an Eötvös balance.

In Weber's scheme, neutrinos coming from essentially any source would interact with the crystal as a whole, through coherent neutrino scattering, rather than a single atom as was the case with existing neutrino detectors.[3]

Although each neutrino in Weber's scheme would deposit a tiny amount of energy and momentum into the sapphire, he believed it should be enough to exert a measurable force on the crystal. The force could indicate the direction the neutrino came from. Two identical crystal-based detectors at different locations would reveal a sub's latitude and longitude through triangulation, and three would be enough to add depth and precisely identify the vessel's position in the ocean.

According to Weber, his system even saw the forces that neutrinos coming from tritium created. All other methods for seeing tritium neutrinos rely on studying the electrons the atoms emit, rather than the neutrinos directly. If Weber was detecting them, it meant he'd managed an astounding sensitivity far beyond the capabilities of any other type of neutrino detector in existence then or now. In experiments using more intense reactor sources, blocking the neutrino flow with another sapphire crystal, Weber claimed, caused the signal to disappear. It provided what Weber believed to be a clear indication that the detector was working.[4]

If true, it was a potential boon for nuclear sub hunters, and a nightmare scenario for nuclear sub crews. It would undermine a crucial branch of the U.S. nuclear weapons strategy that relied in large part on the unpredictability and invisibility of oceangoing nuclear weapon platforms.

Although few physicists believed the claims, military leaders were rattled. Weber was well known both for his gravitational wave antennas and for his seminal work in developing the laser. Records released by the Nobel Prize committee show that Weber was nominated twice for the Physics Nobel for his laser research.[5] Weber was of a stature in physics that the Navy was inclined to be attentive to his efforts, no matter that others in academia doubted its foundations.

The combined intelligence promise and security threat of Weber's detector, and a related scheme proposed by the Raytheon Corporation, deeply concerned the U.S. military. That led a semi-secret team of elite U.S. science advisors known as the JASONs to step in to provide some clarity.[6]

The JASON advisory group is a nebulous organization with a mysterious acronym that may or may not reference Jason and the Argonauts, or perhaps the first letter of each of the months of the year they meet (July through November), depending on whom you ask. The group includes an unspecified number of leading scientists, probably between 30 and 60 members. The JASONs formed in the wake of the Soviet launch of the Sputnik satellite, which led to a crisis of confidence in U.S. technological competence. They have since advised the nation's leaders on critical scientific issues such as particle beam weapons, data mining the human genome, and climate change.

In 1988, they turned their sights on Joe Weber's neutrino detectors. Three members of the JASONs, Princeton University physicists Curt Callan and Sam Treiman, along with Freeman Dyson of the Institute for Advanced Study, spent seven weeks assembling a report. The 97-page document includes plain-language descriptions of coherent neutrino scattering as well as a compilation of what the authors labeled theorems and proofs that they hoped would put the proposals to rest.

"This was a collective effort," wrote Callan in an email recalling the effort. "Treiman and I were professional particle theorists and therefore well-versed in the physics of neutrinos and their detection. Dyson was a mathematical physicist who talked in theorems. I think we probably decided collectively that presenting our results as 'theorems' would help put a definitive stake through the heart of the claims of the more fantastic schemes for neutrino detection that people were trying to sell to the [Department of Defense]."

From the beginning, the JASONs knew there was never much hope that the schemes were viable. "[A]s physicists, we didn't really need formal theorems to reach our main conclusions," wrote Callan. "The goal was to inoculate decision makers against wild and physically incorrect claims about military/security applications of neutrinos."

Callan doesn't recall getting any feedback from the military brass they targeted with the report. "In any event," wrote Callan, "the fantastic claims never got any real traction." Military support waned and neither the Raytheon proposal nor Weber's experiment were ever definitively supported by other experimentalists in the years that followed. Because the report had seemed to serve its intended purpose, the JASON authors never bothered to formally publish it.

"We were hoping to address an audience that doesn't read the scholarly literature, and time was of the essence," wrote Callan. "Perhaps we should have converted our report into a scholarly article, but since we were at some level just stating the obvious (obvious to a good particle physicist anyway) we probably did not feel that our results really merited publication in *Phys Rev* or equivalent. Maybe we were wrong about that, but we were all busy with other things."

According to Trimble, Weber didn't know the scale of the ruckus his detector inspired. "I was not previously aware of the JASON report, and am not sure Joe was either," wrote Trimble in response to an inquiry about the episode.

Even if Weber's detectors could have picked up neutrinos from submarines, they would have suffered from at least one fatal flaw. "Coherent scattering is sensitive to all types of neutrinos, so it's sensitive to solar neutrinos," says neutrino physicist John Learned of the University of Hawaii.[7] "Once you go a couple of kilometers away from a reactor, such a detector, if it did exist, would be only measuring solar neutrinos because they'd be so dominant."

Despite the transient panic that Weber and Raytheon triggered, the earlier top secret report's findings still hold today: Regardless of the method involved, neutrino detectors capable of tracking subs are still much too large and too expensive to replace acoustic listening posts and other forms of surveillance.

Coherent Scattering for Real

It's clear now that Weber's coherent scattering scheme was fatally flawed. Still, there was a seed of truth in his effort. It wasn't the first time Weber was in that sort of position. He had reported the first observation of gravitational waves in the 1960s. Weber was almost certainly wrong then too.

Near the end of his life, Weber felt that he had become an outcast in the scientific community for insisting that his broadly dismissed gravitational wave claims were correct. His legacy has been redeemed with the discovery of gravitational waves at the Laser Interferometer Gravitational Wave Observatory (LIGO), 16 years after Weber's death in 2000. He's now widely regarded in the physics community as the father of gravitational wave astronomy. Coherent neutrino scattering was experimentally confirmed in 2017, the year following LIGO's gravitational wave triumph. Nevertheless, Weber's neutrino efforts are not so fondly remembered.

Neutrinos don't scatter from macroscopic crystals in the way Weber hoped, but they can bounce off the nucleus of an atom. While far less sensitive than crystal detectors would have been if they had worked, under some circumstances the effect offers sensitivity much greater than the gargantuan, fluid-filled tanks that have dominated neutrino research over the last half century. In contrast to kiloton underground tanks and gigaton underwater and ice-embedded arrays, some coherent scattering detectors are the size of a milk jug and weigh about as much as a toddler.

Coherent neutrino-nucleus scattering, which physicists abbreviate as CEvNS and pronounce "sevens," was predicted in 1974 when the Standard Model was young.[8] It was finally confirmed in 2017 in an experimental hall called Neutrino Alley, adjacent to the Spallation Neutron Source at Oak Ridge National Laboratory in Tennessee. CEvNS hinges on the enhanced probability that a neutrino will scatter off a nucleus, under the right experimental conditions. Coherent scattering improves the sensitivity of detectors by hundreds of times and increases with the square of the number of neutrons in a nucleus. A target material with twice the number of neutrons as another material raises the chances of scattering by a factor of four; tripling the number of neutrons increases scattering chances nine-fold; and so on.

It seems simple enough. Bigger nuclei lead to more scattering, which means higher sensitivity to neutrinos and comparatively

A two-kilogram prototype of the 14.5-kilogram CEvNS detector used in Neutrino Alley at Oak Ridge National Laboratory is literally small enough to sit on a desktop. *Source:* Courtesy of the COHERENT Collaboration.

tiny detectors. So, why did it take 40 years to confirm the effect? And why have other detectors grown so large, with still larger ones on the way? The answer to both questions is that the neutrino sources are different from the ones the big detectors have typically relied on.

As you might imagine from the name, the Oak Ridge Spallation[9] Neutron Source was built to make neutrons. They are knocked from mercury atoms by pulses of high-energy protons that slam into a liquid mercury target 60 times per second. The neutrons are cooled and allowed to pass through gaps in shielding to create beams for a variety of research applications, including studying the structure of materials like superconductors, metals, plastics, and biological samples, as well as helping to maintain and monitor materials crucial to U.S. nuclear weapons stockpiles.

Neutrinos, on the other hand, are an unintended byproduct of the spallation neutron source. Some of the energy from the proton collisions creates pion particles. Unlike experiments that produce neutrino beams, a spallation source target is large and the pions rapidly slow down inside the dense material before they decay into an antimuon particle and a muon neutrino. Because the pions are essentially at rest when the decays happen, and only two particles result from each, the muon neutrinos all come out at very nearly the same, low energy.

The uniformly low-energy muon neutrinos emerge in intense bursts promptly after the firing of each pulse of protons onto the mercury target. Despite their intensity, any signal they might produce in a detector immediately adjacent to the source would be swamped by the neutrons that are the intended product of the spallation source. Luckily for the researchers looking for coherent neutrino-nucleus scattering, there's a more accommodating space nearby, thanks to the former utility hall that's now Neutrino Alley.

"That was a great piece of good fortune," says Kate Scholberg, a Duke University professor who is among the pioneers in coherent scattering research and serves as spokesperson for a collaboration known as COHERENT. "There's a corridor that runs along the beamline and it's got some overburden [building material above the basement-level Neutrino Alley] so it gets rid of some cosmic rays. And it's also got a whole bunch of shielding, which we didn't put there on purpose, it was just put there randomly . . . of course there always has to be some shielding for humans to go there, but there's a lot of it and it really completely kills the background neutrons. We can get as close as 16 meters from the neutrino source with almost no neutrons." Neutrino Alley is an excellent place to set up shop, if you need intense bursts of low-energy neutrinos and relatively little background radiation to confound measurements.

Neutrino Alley is a tight space, far too small to accommodate the gargantuan detectors that rely on more established approaches, even given the intense, low-background muon neutrino pulses. Coherent scattering detectors fit just fine. The canister of liquid argon that served as the detector for the first confirmed coherent scattering experiment is small enough to sit on a desktop. The addition of photodetectors, a vacuum chamber, and shielding results in an experiment about meter across and two meters tall. That's small even by the standards of the very first neutrino detectors that Cowan and Reines built.

Lucky CEvNS

In high-energy detection experiments, the neutrinos primarily interact with a single subatomic particle in a nucleus, ramming into a neutron, for example, and creating a detectable shower of particles. Coherent neutrino-nucleus scattering isn't so specific. "It's a kind of scattering where the neutrino just kind of smacks the nucleus as a whole object," said Scholberg. "A high energy neutrino is like a bullet blowing the whole apple up, or you can have a medium energy bullet where it goes through the apple and bits of debris get spit off of it, or you can gently smack the apple and it just kind of rolls off. It's a really high cross section, meaning that it happens a lot by neutrino standards. Neutrinos of course are weakly interacting and so they hardly ever kick anything at all, but with the coherent scattering, you get a lot of them by neutrino standards."

The analogy isn't perfect. As particles that take part in the weak force, they interact by exchanging force particles like the Z and W bosons. The key to coherent scattering is to dial down the energy of the neutrinos. That reduces the energy of their exchange bosons. Thanks to wave-particle duality, the boson wavelength is stretched. The bosons see an atomic nucleus as a whole, instead of seeing individual neutrons in the atoms. As a result, says Scholberg, "The nucleus is just a blob of weak charge, it's just a single coherent object," as far as the boson is concerned.

The muon neutrinos that come from the Oak Ridge source emerge with energy low enough to ensure their force-exchange bosons' wavelengths are comparable to the sizes of nuclei like argon, cesium iodide, and germanium that serve as target materials.

Wavelength tuning isn't the whole story. Otherwise, the scattering would be proportional only to the number of neutrons in the target nucleus, rather than the neutron number squared. This is where coherence plays a crucial part.

Fortunately, coherence is a property that can be understood without resorting to quantum mechanics. For example, the trumpet section in a band might consist of 10 trumpets. Each trumpet emits a sound wave, which you can think of as a series of crests and troughs that impinge on your ear. The energy carried by the sound wave is given by the square of the amplitude of the wave, the amplitude being the difference in height between the top of the crest and the bottom of the trough.

Typically, when the 10 trumpets are all playing the same note, they will not be in phase. That is, the crests and troughs from the different trumpets do not line up in any particularly organized way. Under such circumstances, the 10 trumpets deliver about 10 times the energy to your ear that a single trumpet would.

If by chance all the crests and all the troughs happened to line up coherently, they would produce a gigantic wave with an amplitude 10 times that of the wave from a single trumpet. Because the energy in the wave is the square of the amplitude, your ear would experience a hundred times the energy of one trumpet alone. In the world of musical instruments, this never happens, otherwise classical concerts would be much more painful than they currently are.[10] But in the world of neutrino scattering, coherence does happen. Think of each nucleon as analogous to a quantum-mechanical trumpet that plays a note whose strength gives the probability of its scattering by the Z boson. The total probability is given by adding the sound from each nucleon, and then, just as with ordinary waves, squaring the result. If the Z boson's momentum is small enough that it interacts with the entire nucleus, it stimulates a coherent burst of "sound" from all the nucleons. The neutrons and the protons respond differently to the Z, with the neutrons producing the much stronger sound. Effectively, the scattering probability is proportional to the number of neutrons in the nucleus squared. A typical experiment might use cesium atoms, which contain 78 neutrons. The scattering is about 6,084 times as strong as from a single neutron.

The problem, explains Scholberg, is that even with coherence, the recoils you get from long-wavelength, low-energy interactions are small. "It's like you have a ping pong ball and it hits a bowling ball, and that bowling ball rolls off very gently."

Although it takes place with high probability, relative to noncoherent scattering, it leaves behind only the slightest trace of having happened. The scattered neutrino cannot be directly detected. What is detected is the recoiling nucleus.

The nucleus that is struck is embedded in a scintillating material that produces a small flash of light when the nucleus recoils.

The experiment must be designed so that the flash is detected and distinguished with high certainty from the many background processes that can mimic the signal.

"It's really only been since the mid-2000s or so that there has been technology focused on seeing that tiny recoil," says Scholberg. "In fact, that technology was built for a different purpose. It's been built to look for WIMPs [weakly interacting dark matter particles], to look for dark matter. Since that is exactly the same signature that CEvNS will give you." No WIMPs have clearly turned up in any experiment so far, but if they exist, coherently scattering neutrinos may well be a problem for WIMP hunters. "At some point, you get so sensitive that you get blinded by the CEvNS from the sun . . . that level is called the neutrino floor and that's kind of a limit for how far WIMP dark matter detectors can go."

Neutrinos, which have been so hard to detect in the past, are now the chief source of noise that other scientists must contend with. And because they're neutrinos, there is no amount of shielding that can block them. That is, there are no "WIMP Alleys" that researchers can count on in the foreseeable future.

What Can CEvNS See?

The foremost goal of the COHERENT collaboration at the Spallation Neutron Source, as Scholberg describes it, is to test the Standard Model and, it is hoped, find where it breaks down. "CEvNS was predicted in 1974 in the Standard Model, and it has a quite well understood cross section. You don't have to worry, to first order, about nuclear effects because the nucleus looks like just a blob to the neutrino," says Scholberg. As a result, things that complicate high-energy nuclear physics experiments are not a problem for coherent neutrino scattering.

"That's actually kind of a big deal in nuclear high energy physics, that it's hard to understand the interactions very well because nuclei are big, messy, goopy places," says Scholberg. "Relatively speaking, we know precisely what is going to happen. What that means is that if there's some new physics, if there's some new interaction of a neutrino, then we can tell because we know what we expect from the Standard Model. If we find something that deviates, we know it's from new physics and not some crazy nuclear effect."

The COHERENT collaboration researchers have so far measured the expected rate of neutrino scattering in cesium iodide crystals of the world's smallest working neutrino detector, and in the nuclei of argon atoms. Measurements of the effect in sodium iodide and germanium are in the works. "It's a broad test of the Standard Model to have a whole bunch of different kinds of nuclei, light ones and heavy ones," says Scholberg.

Experiments at nuclear reactors could test the Standard Model more extensively by taking advantage of the copious neutrinos they produce at still lower energies than muon neutrinos that come from the Spallation Neutron Source. The Coherent Neutrino Nucleus Scattering (CONUS) experiment, for example, is positioned 17 meters from the core of a commercial reactor in Brokdorf, Germany. CONUS uses germanium detectors exposed to 24 million neutrinos per square centimeter every second. The Coherent Neutrino Nucleus Interaction Experiment (CONNIE) uses silicon rather than germanium, at the Angra dos Reis nuclear power plant in Rio de Janeiro, Brazil. Both should be up and running in the early 2020s.

Low-energy neutrinos from the sun and supernovas will also turn up in coherent scattering measurements. "In fact, that's where I first heard about this process," says Scholberg, who has dedicated much of the research over her career to astrophysical neutrinos. Next-generation dark matter detectors, like the 40-ton Dark Matter Wimp Search with Liquid Xenon (DARWIN) in Europe, will both keep an eye out for supernovas and study the low-energy solar neutrinos that otherwise provide the noise floor that limits WIMP searches. CEvNS studies of low-energy neutrinos from high-flux sources like reactors and accelerators will also help to clarify the particle dynamics in supernovas, where coherent scattering is an important process.

In Search of Sterile Neutrinos with CEvNS

Coherent scattering could be used to search for sterile neutrinos that have been invoked to explain the curious neutrino anomalies in experiments at LSND and MiniBooNE.

It's entirely possible that researchers don't understand their neutrino sources or detectors well enough, which might account for the anomalies. But some people argue that they imply that there could be a fourth type of neutrino to join the three flavors we know about. No detector of any kind could see sterile neutrinos because they don't interact at all except perhaps through gravity (and their masses are way too small to detect gravitationally in any lab experiment). But one of the best ways to not see sterile neutrinos in the lab might be with coherent scattering detectors.

Coherent scattering detectors are sensitive to all three known neutrino flavors, in both their matter and antimatter guises, because of their interactions through the weak force. The detectors are also small enough that they can be moved about to measure how neutrino populations change over the course of meters. Neutrino Alley is too tight a space to do the experiment. But researchers at the Los Alamos neutron scattering facility are working on an experiment called Coherent CAPTAIN-Mills (CCM)[11] to look for sterile neutrinos or, more specifically, to infer them from the absence of the three known neutrino flavors. They're doing it in the same laboratory where the LSND anomalies first turned up.

As at the Oak Ridge lab, neutrinos at the Los Alamos Neutron Science Center (LANSCE) are byproducts of knocking loose neutrons from the target with pulses of energetic protons. The target at LANSCE is tungsten, rather than the mercury of the Oak Ridge source, but the result is a similar pulse of neutrons, pions, and promptly emitted muon neutrinos. CCM's 10-ton liquid argon detector is much larger than the 22-kilogram tank that COHERENT researchers used to first see coherent neutrino scattering in argon. Unlike the tight quarters in Neutrino Alley, LANSCE is spacious enough that CCM researchers will be able to move their detector from time to time, or install additional, identical liquid argon tanks to simultaneously map neutrino populations and identify sterile neutrinos by the absence of the other three.

The experiment may finally rule out the controversial LSND findings that have dogged neutrino researchers for nearly 30 years, or perhaps add more compelling evidence for light sterile neutrinos that would indicate a crack in the Standard Model. If they're very fortunate, and the properties of sterile neutrinos are just right, the CCM team could find a dearth of neutrinos at one distance, and an increase farther away, as sterile neutrinos oscillate back to the detectable flavors. It would be a dramatic confirmation of the—so far—tenacious anomaly.

More for CEvNS Soon

In only a few years, coherent elastic neutrino-nucleus scattering has gone from an old, unfulfilled Standard Model prediction to Weber's red herring to experimental confirmation to, finally, the core technology for a rapidly expanding list of experiments and potential applications. In a field where advances over the last half century consisted primarily of larger and more refined versions of devices that have been around for decades, coherent elastic neutrino-nucleus scattering has opened new avenues of neutrino research.

CEvNS-based applications are limited to arenas where neutrino energies are low, which means that they add to researchers' tool-boxes but don't replace the more established detectors. While it's unlikely you will ever need a neutrino detector that can fit on your kitchen counter, if you do someday, it appears that it will be one based on coherent scattering.

12

Infinity through a Keyhole

I'm interested in neutrinos because there are so many really basic things we don't know about them.
—Diana Parno, neutrino physicist, Carnegie Mellon University

Much of the mystery surrounding neutrinos hinges on mass. The fact that they oscillate from one flavor to another means that the three mass states are distinct. One of them may be zero, but the other two must be different from zero and from each other. Of course, it is possible that none of the three is zero. It's not yet clear which is the most massive of the three.

Experiments to directly determine neutrino masses have roots that extend back to the very beginning of the neutrino saga. Although it could not have been his intention, considering he didn't know neutrinos existed at the time, Chadwick's beta particle experiments were the first crude neutrino mass measurements. While the beta decay spectrum that Chadwick discovered was initially a source of debate and confusion, it still provided enough information for Pauli to sketch out key neutrino properties, make a very rough estimate of its mass, and take a reasonable guess about its magnetic moment.

A closer look at the shape of the spectrum can provide additional key information about neutrinos. In particular, at the higher energy end, in the range where the beta particle ends up with most of the reaction energy and the neutrino carries off very little, there is likely to be a slight deviation in the energy. That deviation could reveal neutrino mass. One thing we know is that the neutrino mass is very small; as a result, the effect on beta decay is also very small. Discerning the difference requires some extremely precise instruments and challenging experimental techniques.

KATRIN

The Karlsruhe Tritium Neutrino (KATRIN) experiment in Germany is a spectrometer that sorts electrons based on their energy. "What we want to do with KATRIN," says Diana Parno, a Carnegie Mellon University professor who is part of the KATRIN data analysis team, "is to try to measure the neutrino mass with as few assumptions as we can get away with."[1] Other measures of neutrino mass, like estimating it by looking at the influence that neutrinos have on the structure of the cosmos, require factors that aren't directly related to neutrinos. Evolving measurements of the distribution of galaxies in the universe can lead to different neutrino mass estimates,[2] undermining confidence in the approaches. Experiments that indirectly infer the masses can suggest different values depending on the theories that help interpret the results.

The KATRIN experiment is an attempt to get at the neutrino masses more directly by measuring the electron energy in beta decays. You still need to calculate the neutrino mass by studying electrons, but there's no guesswork involved. Thanks to the conservation of energy and momentum, it's easy to understand what's going on with the neutrino from the electron energy.

KATRIN begins with a chamber filled with tritium gas. The half-life for tritium is a little over 12 years. But the KATRIN source is loaded with enough atoms that about 100 billion of them decay to helium each second.

In KATRIN, the spectrometer is roughly the shape of a rugby ball, though much larger at nearly 10 meters in diameter, comparable to the height of a two-story house. It's twice as long as a school bus and weighs almost 30 times as much. The size of the spectrometer, which Parno likens to a blimp, accounts for the spread in directions that the electrons have as they enter the spectrometer. "There's a strong magnet at each end of the spectrometer, like these kind of pinched ends, and then in the center of the blimp where it's really broad there's a region of really weak magnetic field," says Parno. "If you're gradually guiding the electrons between a region of strong magnetic field and a region of weak magnetic field, you're going to rotate the momentum of the electrons. In the middle where the magnetic field is weak, the electrons are going to be moving almost exactly parallel to the magnetic field lines . . . we go through all of this to get these electrons all moving in more or less the same direction."

KATRIN doesn't measure the energy of the electrons directly. Instead, it includes an electric field that slows the electrons down as they make their way through the spectrometer. "An electron is going to need a certain amount of kinetic energy to pass this barrier," says Parno. "If it has less energy than the threshold then it will be reflected back toward the source, if it has more energy it will transmit through."

It's a little like you and a group of friends trying to coast over a hill on your bicycles (assuming you all agree not to pedal after you've come up to speed). If you have enough energy, you can glide up and over a small hill easily. The slower among you may not be able to coast all the way to the top, others will roll on over, depending on how much energy each starts with. For progressively higher hills, fewer will make it. By counting the bicyclists that can clear hills of different sizes you get a measure of the distribution of energy among the group. When you find the hill that is just big

The KATRIN spectrometer on the way through Eggenstein-Leopoldshafen in Germany in 2006, destined for installation at the Karlsruhe Institute of Technology. *Source:* Courtesy of the Karlsruhe Institute of Technology.

enough to stop anyone from coasting over, you will have a handle on the highest energy of the cyclists in the group. In KATRIN, turning up the electric field until no electrons pass allows researchers to calculate an upper limit on the electron energy, and in turn narrow in on neutrino mass.

Only about one in 12 trillion of the electrons that enter the KATRIN spectrometer is in the energy range that can supply information on the neutrino mass. The mass of neutrinos that accompany beta decays affects the electrons with the highest energies coming out of the tritium source. At the end of the planned run for experiment, KATRIN researchers expect to see sensitivity down to neutrino masses about 25 million times lighter than the electron, the next heavier particle in the Standard Model. If neutrinos are any lighter, KATRIN will offer only an upper limit on neutrino mass.

"We haven't discovered a non-zero neutrino mass . . . everything we've discovered has been consistent with zero so far. Of all the data we've analyzed, the true neutrino mass lies somewhere smaller than our sensitivity," says Parno. If nature is unkind to KATRIN, and the neutrino mass that the collaboration is after is too small for the spectrometer to see, there are other neutrino questions for them to pursue. "After we reach the [design] sensitivity, one really exciting idea that we're working on is to upgrade the detector and make some changes to other parts of the apparatus so that we can extend our highly detailed energy spectral measurement much, much deeper into the tritium spectrum, so that we could handle higher rates at lower beta energy. That would allow us to look for sterile neutrinos," which might be thousands of times the mass of the neutrinos KATRIN researchers are currently seeking. "At that point," says Parno, "it's actually a candidate for dark matter."

Project 8

One of the challenges that KATRIN researchers have to contend with is the added complication from the tritium itself. As a form of hydrogen, tritium is very reactive. It readily forms molecules with other atoms. To keep their tritium supply stable, KATRIN uses a tritium gas where pairs of tritium atoms have already combined to create stable, less reactive molecules each composed of two tritium atoms. While necessary to maintain the tritium gas supply, it complicates the beta decay measurements. Molecular pairs of atoms can spin or vibrate in ways that smear out the energy of an emitted electron. Imagine a pair of billiard balls connected by a rubber band—they can move in ways that individual billiard balls couldn't. "They're small effects," says Parno, "but we're looking for a small change. We have some really beautiful molecular theory that gives us a precise description of what these molecular effects are, and we simultaneously look for experimental tools to make sure the theory is correct."

Another planned beta decay experiment could potentially do five times better than KATRIN, in part by relying on individual tritium atoms instead of molecules. The experimenters building Project 8 plan to use magnetic fields to confine tritium atoms to keep them away from the walls of their experiment and prevent them from reacting with other materials or each other. "For Project 8," says Joseph Formaggio, the experiment's spokesperson and cofounder, "we already need the magnetic fields anyway, so it's sort of a natural environment for us to operate in."[3] Like KATRIN, Project 8 also involves beta decay electrons spiraling around in a magnetic field. Unlike KATRIN, the magnetic fields will ultimately serve the dual purpose of keeping reacting tritium atoms in check while also measuring beta electron energies.

Instead of using the field to guide the electrons, Project 8 measures the spirals themselves. When an electron moves across the direction that a magnetic field points, it performs circular motions known as cyclotron orbits. For low-energy electrons, the cyclotron orbit frequency is constant, depending on the electron mass, the charge, and the strength of the magnetic field. At high energies, though, the speed of the electrons becomes an appreciable fraction of the speed of light. That is when relativity starts to be important. The frequency of the orbits for the relativistic electrons depends on the electron energy and provides similar information that the KATRIN spectrometer offers.

The Project 8 apparatus will be a good deal more compact than the massive KATRIN experiment. The primary components include a superconducting magnet surrounding a cell that is supplied with a flow of tritium gas. The electron emitted from a decaying tritium atom begins cyclotron orbits in the magnetic field. Any time an electron moves in a curved path, it radiates energy that can be picked up with an antenna.

Instead of creating an electrostatic hill for electrons to climb over, Project 8 tunes into the signals the electrons broadcast as they spiral. The emissions sap energy from the electron, but by recording the initial signal frequency the electron puts out, the Project 8 team can determine the energy the beta emission electron started with. The approach has some inherent advantages that help reduce background noise. As Formaggio puts it, Project 8 operates in *frequency space*, which is essentially what you do when you tune your radio from one station to another. "On your radio station," says Formaggio, "95.5 and 100.3 never mix with each other. You never hear a station that's way down the dial" from the one you're tuned to, "because frequency is ultra-precise in that way." Lower-energy electrons and many other sources of noise simply don't turn up because they're on a different station, so to speak.

Formaggio hopes that Project 8 will be able to measure energies one fifth of the KATRIN lower limit. If so, it increases the chance that the Project 8 team will pin down the mass of neutrinos released in beta decays. In addition, the sensitivity is good enough that even short of supplying a concrete mass measurement, Project 8 may provide clues about the ordering of the neutrino masses.

One other difference between Project 8 and KATRIN, and many other experimental collaborations, is the origin of the name, which initially puzzles many people in the field. "That is by far my most common question when I give talks," says Formaggio. "It's somewhat of a rebellious name. Before Ben Monreal and I conceived of the idea of Project 8, we were just sitting around and complaining about how all these experiments have these really contrived acronyms." In 2010 or 2011, as Formaggio recalls, they vowed to pick a meaningless name if they ever were to invent their own experiment. They decided that "Project 8" would be a good choice. When Formaggio and Monreal later came up with their novel neutrino mass measurement, the question of a name arose. "Ben turns to me and says, 'well obviously we should name it Project 8.' So that's how it got its name," recalled Formaggio—although skateboarder Tony Hawk had happened on the same phrase in 2006 for an installment of his skateboarding video game. "So far we seem to have held onto the name."

Beta Decay in Reverse

The reverse of beta decay, beta capture, can also reveal neutrino mass. The Electron Capture Holmium (ECHo) experiment involves a supply of the rare earth element holmium.[4] An unstable isotope of holmium (^{163}Ho) decays by absorbing one of its own electrons to form another rare earth called dysprosium. Initially, the dysprosium atom is in an energetically excited state. It then emits energy that creates a tiny, but measurable rise in temperature.

Every holmium decay event leads to energy deposited in the absorber, while only the very highest energy decays can offer information about neutrino mass. That means that only one in a trillion of the measurements are useful. By creating enormous arrays of minute and sensitive temperature sensors, the researchers should be able to pick out the important events.

ECHo is in earlier stages than KATRIN and Project 8, but models suggest it could ultimately provide comparable measures of neutrino mass with arrays of tens of thousands of temperature sensors. It could have better resolution, given the larger number of sensors. One strength of calorimetric designs, as heat-sensing systems like ECHo are known, is an easier route to expansion and improved sensitivity, while KATRIN and Project 8 have ultimate resolution that would be difficult to improve without fundamental redesigns.

Neutrinoless Double Beta Decay

If neutrinos have mass, as oscillations suggest and as direct mass measurement experiments seek to confirm, we may soon have experimental evidence to tell whether they are of the Dirac variety, and the neutrino and antineutrino are separate particles like the electron and the positron, or the neutrino is of the Majorana variety and is its own antiparticle.

One experiment might directly determine the Dirac or Majorana nature of neutrinos and solve some of the particles' deepest mysteries. It's a search for a rare nuclear reaction known as *neutrinoless double beta decay*, which physicists sometimes write as $0\nu\beta\beta$ for short (indicating zero neutrinos and two beta particles are involved).

Normal beta decay occurs when a neutron inside a nucleus converts to a proton, emitting an electron and an antineutrino in the process. There are a few nuclei that don't transform via single beta decay but can still decay with the emission of two electrons at the same time. This is called double beta decay.

In the event of single beta decay, one of the neutrons will convert to a proton in a nucleus, changing the initial atom to the element one place farther along on the periodic table. For example, a form of potassium atom with 11 protons and 11 neutrons, on beta decay, turns into magnesium with 12 protons and 10 neutrons.

If the combined mass of a daughter nucleus plus an electron is heavier than the parent, then single beta decay cannot occur. It would result in an increase of mass and energy, with no outside source to supply it. That would violate conservation of mass and energy and, therefore, can't happen.

It can happen, however, that a nucleus with two more protons, and two fewer neutrons, is sufficiently lighter than the parent to allow double beta decay to take place. Because the final products weigh less than the initial atom, the difference in mass and energy is carried away by the particles that come out in a decay. In the case of a xenon atom that begins with 54 protons and 82 neutrons, the chemical result of double beta decay is a barium atom with 56 protons and 80 neutrons.

If the neutrino is a Dirac particle, that's the end of the story. The electrons are matter particles, and the antineutrinos are antimatter particles. Two of each come out of double beta decay and the net number of matter and antimatter particles in the universe is the same. The two matter particles (electrons) and two antimatter particles (antineutrinos) offset each other. The relative amount of matter and antimatter is unchanged.

But if they are Majorana neutrinos, each antineutrino is also partly neutrino. When the two particles meet, there will be some probability that the antineutrino component of one of them will annihilate with the neutrino component of the other. In that case, no neutrino will emerge from the reaction at all, leaving only the daughter nucleus and the two electrons in the final state. This is neutrinoless double beta decay.

Because there are more matter particles after the neutrinoless double beta decay, but no offsetting antimatter particles, it results in a net increase in matter in the universe. If it ever happens, it is a possible key to understanding why there's any matter left in the universe at all.

Decay processes are normally characterized by their half-lives, which is the time it would take for half the atoms in a sample to decay. The weak interaction is responsible for both beta decay and double beta decay. Decays that are due to the weak force generally take a long time. The neutron itself beta decays in about 15 minutes; tritium, which is a heavy version of hydrogen, has a beta decay lifetime of 12 years. Double beta decay involves two simultaneous weak force events, leading to extremely long lifetimes. The double beta decay lifetime of xenon 136 is 100 billion times the age of the universe.

Observation of neutrinoless double beta decay would conclusively prove that neutrinos are their own antimatter particles, as Majorana proposed. Failure to see the reaction implies that the neutrino is a Dirac particle that comes in distinct matter and antimatter versions. But the failure could also be blamed on the smallness of the neutrino mass, pushing the lifetime of neutrinoless double beta decay beyond the reach of current experimental capabilities.

The lifetime of the neutrinoless decay is undoubtedly very long. Still, experiments are closing in on measurement sensitivities where observations of the events seem likely, and the failure to see the decays could tilt the scale toward Dirac neutrinos, or perhaps suggest that there's still more to learn about physics beyond what is described in the Standard Model.

The Monumental Quest to Find Nothing

Neutrinos are hard to catch under the best of circumstances. Determining when they aren't there is even harder. As a result, experimentalists who are searching for neutrinoless double beta decay are dealing with some particularly challenging measurements.

Typical double beta decay, with atoms that emit two neutrinos and a pair of electrons (or the electron's antimatter partners, the positrons), have lifetimes on the order of 100 million trillion years. The age of the universe is only a bit more than 13 billion years, so there is almost no chance that an atom will undergo double beta decay during the course of a laboratory experiment that lasts a few years. The lifetime of an atom undergoing neutrinoless double beta decay is 10,000 times longer still, if not more.

Physicists overcome this by hoping to detect a small number of decays in a very large collection of atoms. While the chances of one atom undergoing double beta decay (along with a pair of neutrinos) in a given year vanishingly small, the chances of seeing a single beta decay per year in an experiment involving very many atoms can quite good. Considering there are trillions of trillions of atoms in a gram of germanium 76, which is one material that can exhibit double beta decay, it doesn't take a large experiment to see the decays in the lab. In fact, several examples of double beta decay have been observed.

So far, though, no one has definitively seen double beta decay without neutrinos. The lifetime is expected to be no shorter than 100,000 billion times the age of the universe, and perhaps much longer.

In addition, neutrinoless double beta decay experiments need to deal with a conundrum: Even in ordinary double beta decay, the two neutrinos escape the apparatus and are not detected. So how is it possible to distinguish the case when the neutrinos are not even produced?

The answer lies in conservation of energy and momentum. It's the reverse of the situation that Pauli faced when he proposed the neutrino in the first place. When single beta decay occurs, if an atom emitted only an electron or positron, then conservation of energy and momentum forces the emitted particle to have a single, well-defined energy. Instead, the beta decay results in a continuous spectrum of the energy, which in turn means some of the energy and momentum were disappearing in the process. It's to account for the apparently missing energy and momentum in beta decay that Pauli invented the idea of neutrinos.

The final state in double beta decay has even more components than in the single beta decay case: the daughter nucleus, two electrons, and two neutrinos. The energy and momentum are distributed among all the particles, which endows the electrons with energy in a spectrum of values. If the neutrinos happen to take up most of the energy available in a decay, the electrons have little left to share. When the two electrons have their maximum energy, the neutrinos carry off zero energy. For most decays, the distribution of energy lies somewhere between the two extreme cases.

For neutrinoless double beta decay, the neutrinos don't account for any of the energy simply because they don't even exist outside the nucleus. All the neutrinoless events are piled up close to a single energy. Detecting neutrinoless double beta decay demands an experiment with very good energy resolution that can distinguish between the relatively abundant double beta decays at lower energy and the neutrinoless decays that come with only high energy.

Experiments that search for neutrinoless double beta decay consist of containers of one of the materials that could decay via the reaction. In some cases, it's a chunk of solid, as in experiments that measure decays in germanium. Other experiments use liquid or gas.

In order to search for neutrinoless double beta decay, physicists compile a sample rich in isotopes susceptible to double beta decay, then measure the energy of the electrons produced with each decay. Even a few kilograms of isotopes of germanium, xenon, or one of other elements is enough to produce detectable levels of double beta decay with an accompanying pair of neutrinos. Finding the neutrinoless form will likely require a ton or more of an isotope and observations lasting years to decades, assuming nature cooperates.

To see phenomena that occur a handful of times over decades in a ton or more of sample material, it's crucial to exclude false positives. They can arise from outside sources, like cosmic rays, contamination of the experimental equipment, and natural radiation in the isotope sample itself. These are in addition to all the measurement errors and uncertainties that accompany any experiment. Shielding is important, leading to experiments located deep underground and surrounded by materials to absorb as much of the confounding radiation that makes it through or originates from nearby contaminants in the Earth.

The experimental search that seems to be the best prospect in the search for neutrinoless double beta decay is being developed under the auspices of the Large Enriched Germanium Experiment for Neutrinoless Double Beta Decay (LEGEND) Collaboration. The experiment is planned to include a metric ton of germanium largely consisting of the variety with 44 neutrons and 32 protons in its nucleus (^{76}Ge). A smaller version of the experiment, named the Majorana Demonstrator in honor of Ettore Majorana's inspirational work, has been running as a proof of principle since 2014 with a smaller, 40-kilogram supply of germanium. As expected, the diminutive Demonstrator has shown no signs of the neutrinoless decay. But it is key to testing the systems that will make up the 25-times larger, ton-scale LEGEND experiment.

University of Texas at Arlington physicist David Nygren and his collaborators are hoping to run an experiment called Neutrino Experiment with a Xenon Time Projection Chamber (NEXT) that adds previously unexploited methods to exclude false positives. The first involves what Nygren calls "spaghetti and meatballs." In the event of a double beta decay, the electrons in a container full of xenon gas fly away from the decaying atom, knocking electrons from other gas atoms along the way. As the emitted electrons slow, the chance of them interacting with the electrons in xenon gas atoms rapidly increases. The result is a pair of tortuous, spaghetti-like paths that end in relatively intense bursts that Nygren likens to meatballs. The spaghetti-and meatballs signal will help them to eliminate background noise that mimics neutrinoless decays but lacks the pasta-like track.

Catching daughter particles immediately after a neutrinoless double beta decay offers another way to discriminate between the real and imposter events. Whether neutrinos are emitted or not, a xenon atom that decays by double beta decay emits two electrons and converts to a barium atom. Because the initially neutral atom has lost two negatively charged electrons, the resulting daughter particle is positively charged. It's then a matter of sweeping up the electrically charged particles from the otherwise neutral gas in the second or so after a double beta decay event. The three factors—twin spaghetti-and-meatball tracks; seeing doubly charged daughter atoms to ensure a couple of beta decays occurred; and precisely measuring the energy of the resulting electrons to determine whether neutrinos were involved—should help reduce the likelihood of false positives to levels low enough to catch neutrinoless double beta decays, if they occur.

Even if Ettore Majorana was correct, and neutrinos are their own antiparticles, it's possible they have properties that make neutrinoless double beta decay rarer than the best current estimates. That would leave the efforts of LEGEND or other experiments unrewarded. It could then take hundreds or thousands of tons of an isotope to confirm the decays in a decade of observations. Such enormous projects might be beyond the price anyone could afford.

For now, scientists are hoping for the best, while designing and building ambitious, though doable, ton-scale experiments like LEGEND. With some luck, an answer one way or the other may be forthcoming in the next few decades.

"The neutrino is a unique particle in that it has that one feature that can tell us about physics at an energy scale unimaginably higher than what we can reach otherwise," says David Nygren.[5] "It's looking through the keyhole at infinity, so to speak." The potential revelations that the particles promise is what inspired Nygren to join the ranks of neutrino researchers. "It's the only really profound physics issue that I know of that we can address. It's the most profound issue because it also has this big connection to the origin of the universe as we know it."

13

Into the Unknown

Here be dragons.
> —Legend appearing on a sixteenth-century globe

Once upon a time, neutrinos were just what physicists ordered: three flavors to fill out the three generations of Standard Model particles. Each of the electrically charged and massive particles in the Standard model—electron, muon, and tau—came with a matching neutrino. The model demands that the neutrinos have no charge and no mass.

This Goldilocks tale didn't last long.

Neutrino oscillations imply neutrino mass, which ensures neutrinos aren't the "just right" pieces of the particle puzzle that they initially seemed to be. Neutrino mass is an inescapable sign that the Standard Model is incomplete as it stands. The multiple outstanding questions about neutrinos, many of which are the subjects of current or planned experiments, could uncover still more challenges to our understanding of the universe and the rules that govern it.

Some of the most intriguing signs that neutrinos imply of physics beyond the Standard Model are, to say the least, long shots. While they're not yet excluded by experiments, the imaginative solutions they suggest to various physics puzzles, and in some cases challenges to basic scientific tenets, means it's best to consider them with a healthy amount of skepticism. On the other hand, it wasn't terribly long ago that neutrino oscillations, masses, and even the particle's very existence seemed similarly outlandish.

Faster than a Speeding Photon

The neutrino is probably the leading candidate for the most mysterious particle known to exist. As an illustration of how bizarre the neutrino may be, it could even violate the axiom that nothing is faster than light. The neutrino could be a so-called tachyon, a particle traveling at greater than light speed.[1]

A faster-than-light neutrino could have dramatic consequences beyond posing a direct challenge to special relativity. According to relativity, there is no absolute way to fix the order of events connected by a particle traveling faster than light, so ordinary cause and effect might no longer make sense. Also, a particle decay that is forbidden by energy conservation when the particle is at rest could occur when the particle is moving fast enough, if one of the decay products is a tachyon.

That means that there are different outcomes in different frames of reference moving at different speeds. It's as if a person running through a museum was to see a priceless vase fall from a pedestal and shatter, while the security guard standing next to the exhibit sees no damage. In our normal experience, the vase either breaks or it doesn't, regardless of how fast each person in the museum is moving. If tachyons exist, it seems that reality breaks down and there's no way to decide which experience is true, the guard's or the speedy museum visitor's.

In the same way, it's only logical that we should all agree whether or not a particle decay happens. After all, we would be looking at a single event. Neutrinos, or anything else moving faster than light, would potentially lead us to see different outcomes while all looking at the same experiment, depending only on our relative motion. It would shake the foundations of what we think of as reality.

From an experimental standpoint, there are two complementary ways to search for tachyonic behavior. The first is very direct: Just measure the velocity. That is, produce a neutrino at one point, detect it a certain distance away, and divide the distance by the time taken to travel from the first place to the second.

The second way involves the mass of the neutrino. Relativity provides a simple equation relating the energy of a particle to its momentum and mass. If the particle travels more slowly than light, its mass is greater than zero. If the particle travels at light speed, like the photon, its mass is zero. If the particle travels at speeds greater than light, the square of its mass is negative.[2]

A measurement of the square of the neutrino mass could determine whether or not it would move faster than light. Neutrino oscillations give measurements of the differences of the squares of the masses of the three neutrinos. But the differences don't tell us whether some, or all, of the individual masses squared are positive or negative. For that we need a measurement that looks at other combinations of the neutrino masses.

The Oscillation Project with Emulsion-Tracking Apparatus (OPERA) experiment measured neutrino speed in experiments beginning in 2009. OPERA was a neutrino detector in the Gran Sasso Laboratory, deep under a mountain in Italy about 730 kilometers from the CERN laboratory in Geneva. OPERA's main purpose was to study the tau neutrino, but the researchers also had data from muon neutrinos that had been produced at CERN and then traveled to Gran Sasso where they were detected. The production time and detection time were recorded, and the distance between the two laboratories was accurately known using GPS.

The big surprise, when the results of OPERA were announced in September 2011, was that the neutrinos were traveling significantly faster than the speed of light. CERN research director Sergio Bertolucci made the announcement in a webcast conference with invitations extended to members of the news media. The event short-circuited the normal scientific process, which involves peer review that typically provides checks on imprudent announcements and potentially shoddy research. The OPERA measurements showed that neutrinos coming from CERN would have outpaced a photon by about 60 billionths of a second. While this may seem small, it corresponds to a large negative mass squared, and shattered the universal speed limit. If true, it meant neutrinos would have beaten light over the course of the 730-kilometer trip by 18 meters, roughly a fifth the length of a football field. The announcement captured headlines around the world and caused an uproar in the scientific community.[3]

The OPERA press conference unleashed a flood of papers from scientists around the globe. Many offered explanations of how the neutrino could perform this remarkable feat. Others purported to prove that the neutrino could not be a tachyon because of conflicting experimental measurements. Some, such as Boston University's Andrew Cohen and Sheldon Glashow, who published in a paper in *Physical Review Letters*,[4] relied on calculations and theory to throw cold water on the faster-than-light neutrino claims. Cohen and Glashow argued that if neutrinos go faster than the speed of light, they will very quickly lose energy by radiating electron-positron pairs. "Then it's straightforward to calculate the rate," said Glashow, "and the rate is very fast so that the neutrinos . . . could not travel the several hundred kilometers between CERN and the detector. They probably couldn't travel more than a hundred meters, or ten meters," before radiating away their energy. "It was definitely a nail in the coffin," for the OPERA claim, said Glashow, "until the real nail was found. Anyway, they don't go faster than the speed of light."

Yet other papers dwelt on the philosophical difficulties, such as the possibilities of time travel, and the upending of the relation between cause and effect that would be allowed by tachyonic behavior.

Finally, the matter was resolved in the spring of 2012 when OPERA discovered that a loose cable in their apparatus had contaminated their timing data. When the problem was fixed, the discrepancy between the neutrino velocity and the speed of light disappeared (and the wisdom of the conventional peer review process was bolstered).

This did not mean that the neutrino was definitively proved not to be a tachyon. What it did prove is that, if it's a tachyon, its deviation from the speed of light was too small to be measured by OPERA.

The KATRIN experiment, and others that are out to study neutrino masses, could indirectly indicate that neutrinos move faster than light. They don't measure any one of the individual mass states—m_1, m_2, or m_3—but instead reflect a blend of all three. By itself this measurement does not say if any of the masses squared is negative, but if the combination as a whole is negative, then it is assured that at least one of them moves faster than light.

Until recently, both KATRIN and its predecessors had consistently reported negative values for the effective neutrino mass squared. Their results were also consistent with either zero or a small positive value. Most scientists believe negative mass squared is at odds with accepted physics, and therefore impossible. So KATRIN measurements are usually considered as placing a positive upper limit on the effective mass squared, and the negative values are simply ignored.

In 2021, the KATRIN collaborators analyzed more data and reported a small positive value for the electron neutrino's effective mass squared. Once again, uncertainties in the experiment mean the result is consistent with zero or even a small negative value. KATRIN is an extremely difficult experiment, but more accuracy must be gained before it can claim to have actually measured neutrino mass rather than just placing an upper limit on its magnitude. This can be achieved both by analyzing more data and by refining the experiment itself.

The team publishes data that seem to allow the possibility that the squared neutrino masses might be negative because their measurements are so close to zero. As KATRIN researcher Diana Parno explains, they allow for the imaginary mass in the study for mathematical convenience. "If you require your fit to never give you a negative number, you're going to end up with a fit that misbehaves, because you're cutting off half the allowed fluctuation range. And your fit is going to return the number that's just at the edge of the space you've artificially allowed it, but it's not going to give a trustworthy result," says Parno. "We certainly don't expect it to be negative."

Are the philosophical arguments against faster-than-light tachyons sufficient to rule them out? A vote within the physics community would probably overwhelmingly discount the possibility that anything can outrun light. But the history of physics suggests caution. If you're going to make assumptions about the properties of the physical world, you should be prepared for experiments to disagree. Classical physics was upended when quantum mechanics came along. The logical expectation that nature does not distinguish between left and right turned out not to be true of the weak interaction. If tachyons do indeed exist, nature will tell us how to resolve what seem to be paradoxes involving time travel and causality.

The Search for a Special Place in the Universe

Even if KATRIN, or perhaps a more sensitive successor experiment, finds that neutrinos are not tachyons, there is another effect it could check that's outside the boundaries of contemporary physics:

the principle that no place in the universe is more special than any other. More precisely, *Lorentz invariance* asserts that the laws of physics are the same everywhere, regardless of location, orientation, or relative motion.

The earliest tests of Lorentz invariance went in search of a mysterious substance known as the luminiferous ether. Experiments had shown that light had wave-like properties much like sound in air and waves in water. It seemed only logical at the time that if sound and water waves formed ripples in the materials they traveled through, then light needed something to travel through as well. Because light comes to us from faraway stars, the luminiferous ether has to extend through all of space at least out to the most distant discernible parts of the universe. If that was true, the Earth must be gliding through the ether as it moved around the sun. This in turn meant that measuring the speed of light in different directions should reveal the ether's existence because it would cause light to travel faster toward us from one direction than it does from others.

Albert Michelson and Edward Morley built an experiment in the late 1880s to compare light traveling in perpendicular directions. For the time, it was extraordinarily sensitive. Despite isolating their apparatus deep in a basement on the campus of what is now Case Western Reserve University in Cleveland, even the footfalls of passing horses were enough to disturb their measurements.

Ultimately, Michelson and Morley saw no sign of the luminiferous ether. They found the same light speed no matter where they pointed their experiment. It was a major step toward showing that light didn't need a medium to propagate and, more importantly, that light traveled at the same speed relative to any observer, regardless of how they were moving or where they were located. The effort is often touted as the most important failed experiment in history.[5] It sowed the seeds of Einstein's special theory of relativity, which hinges on the principle of Lorentz invariance that states that there is no special location or direction in the universe.

Increasingly sensitive experiments over the last century have found no sign of a breakdown in Lorentz invariance, but there's good reason to think that there is a limit to the principle. We know the Standard Model isn't complete. It doesn't include gravity, which may seem like a major oversight to those of us subject to its perennial pull. To miss such an important piece means it's likely that the Standard Model is just an approximation of a broader theory that takes gravity into account. There are hints that including gravity might introduce violations of Lorentz invariance. If so, whatever theory that wraps up gravity with the particles and forces of the Standard Model would have to violate Lorentz invariance too.

A theory known as the Standard Model Extension (SME) is explicitly designed to incorporate a wide variety of ways in which Lorentz invariance violation might occur, including effects that involve neutrinos.[6]

If there were a special place and direction in the universe, then as the planet rotates on its axis and orbits the sun, we would move relative to that place and direction. That, in turn, would affect the outcome of neutrino experiments.

Characteristics of neutrinos that change in coordination with the Earth's motion would be confirmations of Lorentz invariance violations. Many experiments hunt for the violations by studying the fluctuations in the rate of neutrino oscillations. Neutrino telescopes like IceCube and ANTARES, reactor studies including Daya Bay and Double Chooz, and accelerator-based experiments like MiniBooNE and T2K have all contributed to the search for Lorentz invariance violations in neutrino oscillations, to no avail as yet.

The spreading of neutrinos as they travel from distant astronomical sources could reveal the violation in neutrino telescopes.[7] A breakdown in Lorentz invariance might also lead to different energies carried away by neutrinos in double beta decay studies. In those cases, the violations are independent of the Earth's motion.

KATRIN is poised to provide one of the most sensitive neutrino tests of Lorentz violation so far. The motion of the Earth relative to the preferred reference frame, if it exists, may result in fluctuating measurements of neutrino masses in labs on the planet. If the masses seem to change in harmony with the Earth's motion, it would be an indication of Lorentz invariance violation. That is, if the measurements cycle through different values in ways that repeat over the course of a day, or are in sync with the seasons, it would suggest there is a special place and direction in the universe. It would also make it clear that the special place is not where we are. This would be a compelling sign of Lorentz invariance violation and of physics beyond the Standard Model. It's something the group will keep a skeptical eye out for. "I think if we found it, we would be very surprised," said Diana Parno.

Portal to the Dark Sector

We know from studying the motions of stars around galaxies and the accelerating expansion of the universe that most of existence is fundamentally invisible to us, with roughly 68 percent made of dark energy and 27 percent made of dark matter. It's not something that's part of the Standard Model of particle physics, which is one sign that the esteemed theoretical framework is incomplete, despite its extraordinary successes. Sterile neutrinos, if they exist, would also be beyond the Standard Model. Some theorists have suggested that the two may, in fact, be linked. Sterile neutrinos and dark matter may interact in the *dark sector*, an effectively invisible parallel to the world in which we live.

If the three neutrino mass flavors are joined by more sterile neutrinos that participate in neutrino oscillations, and the sterile neutrinos interact with dark matter, they may give us a way to probe dark matter. In that case, neutrinos potentially offer a portal to the dark sector. We could use the portal to interact with the bulk of matter in the universe that is essentially hidden from us today. The interaction would have the added benefit of explaining the nagging appearance of extra electron neutrinos in experiments at the LSND detector and MiniBooNE at Fermilab.

Some physicists have predicted that the interactions that produce neutrino beams at particle accelerators could also create accompanying beams of dark matter particles.[8] If so, it's likely that the dark matter would be difficult to detect due to an overwhelming signal from energetic neutrinos. One solution is to place a dense obstruction in the beam line that would stop and absorb most of the pions before they have a chance to decay into neutrinos.

It won't stop the neutrinos—nothing stops neutrinos effectively—but it would ensure they have lower energy and are emitted in more random directions, rather than along the beam direction. This would significantly reduce the quantity of neutrinos accompanying a potential dark matter beam. If the dark matter beam exists at all, it would be crucial to filter out the neutrinos that would otherwise swamp a dark matter signal.

Cracks in the Cottage

Goldilocks was disappointed with the too-hot and too-cold porridges she found in the Bears' cottage. But it implied that there might be a just-right bowl waiting somewhere. It's clear now that neutrinos aren't perfect pieces to fill in the Standard Model. Instead, they hint that the Standard Model cottage isn't what it once seemed, and reveal cracks in the walls that could lead to a more glorious structure on the other side. All that's certain is that there is something interesting going on, thanks to neutrinos that aren't just right for the Standard Model.

14

What Can You Do with a Neutrino?

I don't say that the neutrino is going to be a practical thing, but it has been a time-honored pattern that science leads and then technology comes along, and then, put together, these things make an enormous difference in how we live.

—Frederick Reines, neutrino co-discoverer

We often solve problems with the tools available, and not necessarily the best tool for the job. "If the only tool you have is a hammer, you tend to see every problem as a nail," as noted psychologist Abraham Maslow put it. If, on the other hand, you're hammerless but surrounded by nails in need of driving, you might try hitting them with the butt of a screwdriver instead (it's not a great idea). Or you may not notice them at all until you acquire a hammer.

There's no question that there are tasks for which hammers are ideal. In the same way, the unique properties of neutrinos mean they may be the perfect tool for some challenges we now face. Perhaps they will be ideal for other problems that we don't yet realize we have.

If Reines were around today, he might not see neutrinos as practical just yet. But we're getting closer to the time that they will be.

Nuclear Security

Neutrinos are now on the verge of helping to catch nations secretly violating nuclear nonproliferation agreements. It's a natural application that has its roots in the very first successful neutrino experiments. Cowan and Reines relied on comparisons of neutrino flux coming from a reactor when it was on, and again when it was off, to ensure that the signals they were measuring were real and resulted from actual nuclear reactions, rather than some other source of noise. In effect, the flux acted like a neutrino beacon indicating whether the reactor was operating. For physicists like Patrick Huber of Virginia Tech who are developing neutrino-based applications, the approach was prophetic.

"They developed a method to detect neutrinos coming from nuclear reactors which at the time belonged to the Atomic Energy Commission and made plutonium and tritium for the U.S. nuclear weapon program. But you can turn this around," says Huber.[1] "Instead of using reactors to study neutrinos, we can use neutrinos to study reactors."

Huber envisions a portable system that can be quickly and easily deployed to a reactor and provide much more than simple on/off status information. "The number of neutrinos," says Huber, "tells you roughly how much power the reactor is producing. And it's the energy spectrum of the neutrinos which tells you whether the fission that's happening happens to be uranium or plutonium."

Because it's a primary component in nuclear weapons, it's important to keep track of reactor plutonium, particularly for monitoring weapons programs in places such as Iran and North Korea where getting direct access inside reactors can be challenging.

Researchers are close to developing neutrino detectors to monitor facilities such as the Arak Nuclear Plant in Iran for signs of plutonium production for nuclear weapons. *Source:* Nanking2012, CC BY-SA 3.0, Wikimedia Commons (https://commons .wikimedia.org/wiki/File:Arak_Heavy_Water4.JPG).

The detector that Cowan and Reines built, with a ton and a half of scintillator fluid surrounded with lead and paraffin blocks to shield out gamma rays and neutrons, would be too large for the purposes of people like Huber. Even in principle, the technology can't be miniaturized enough to make it feasible for portable systems, primarily due to the need for shielding. New designs rely on additional information to discriminate between neutrinos and other particles.

"What these modern detectors add to the game is also to use the spatial coincidence because the positron and the neutron happen close in space," says Huber. "With modern detectors, we can resolve the position of these two events with good precision at the level of a few centimeters. In the Cowan and Reines days, they looked for the neutron everywhere else in the detector. We just look for the neutron very close to where the positron was, and that gives us an enormous factor of background suppression." In order to do that, explains Huber, they divide the detector into segments, and ignore the signals that are too far apart to result from a neutrino-related event.

"Most cars have a working fuel gauge, and neutrinos for a reactor in a sense are the same thing," says Huber. When operators and regulators monitor reactors today, as Huber sees it, it's comparable to driving a car without a fuel gauge. "Every fueling you carefully record how much fuel you're putting in, then you put the lid back on the reactor, you apply some seals to the reactor . . . and then you run the reactor until it needs refueling. You open up the lid and see whether everything is still there as you put in. And based on the information about how the reactor was run . . . you can figure out how much plutonium should be in the spent fuel. But of course, this is a long chain of inferences and in particular assumes you have access to each refueling, and that between the fuelings nothing else happens."

If you share your car with your roommate, and that person goes to the gas station, then your mileage calculations will be off, because someone put gas in the tank, and you didn't know. And while it might be a pleasant surprise for you to discover extra fuel in the tank, in reactors it makes it difficult to tell if nuclear weapons material has been surreptitiously spirited away between refuelings.

"With neutrinos it's like having a fuel gauge. You can see the amount of plutonium in the reactor core at any given time, and you're not relying on the data from the refuelings," says Huber. "That's attractive in countries like North Korea or Iran where we don't yet have a trustful relationship for safeguards because it gives us the assurance that everything is going well, and it gives the host countries the assurance that we cannot fabricate incidents, because it's very hard to cheat with neutrinos."

Diplomatic relations may not always allow inspectors to get close enough to a nuclear facility for such small-scale monitors to be effective, but neutrinos could still be useful even from afar. For long-distance monitoring, though, as muscle car enthusiasts say, there's no replacement for displacement.[2] That is, detectors have to be large to catch neutrinos coming from reactors more than a few meters distant.

The first neutrino detector specifically intended to demonstrate technology to remotely monitor plutonium production in reactors is the Water Cherenkov Monitor for Antineutrinos (WATCHMAN). The plan is to place WATCHMAN 1,000 meters underground in the Boulby mine in North Yorkshire, England, where it will look for neutrinos coming from the Hartlepool Nuclear Power Station 25 kilometers away. Because of the long distance between the detector and the reactor, WATCHMAN will be large: a cylinder holding as much water as an Olympic-sized swimming pool.

Potentially, detectors could be made very small by exploiting coherent scattering. They are sensitive to all neutrinos, including both antineutrinos that come from reactors and the matter neutrinos that come from the sun. At any significant distance from a terrestrial source, solar neutrinos far outnumber everything else, meaning coherent neutrino scattering isn't feasible, even in principle for remote reactor monitoring. Up close, they can't beat conventional detectors of the type Cowan and Reines pioneered.[3] But there is at least one security-related application where Huber suspects they could have a place: keeping tabs on nuclear waste.[4]

"It seems that for spent nuclear fuel this coherent reaction gives you event rates [very much] larger than what we normally use," said Huber. "Our preliminary results indicate that maybe a 10-kilogram detector could effectively verify the contents of a dry storage cask." The casks are containers that are filled with waste from reactors and sealed. Although it's rare, seals sometimes fail.

"It's not a seal against leaks, it's a seal to verify that the contents haven't been tampered with. And right now, we have no good technology to re-verify the cask contents short of taking it back to the spent fuel pond and opening it up." Huber notes that coherent scattering applications are far behind other neutrino detection methods, but if it pans out the technique could confirm the contents of a storage cask over a period of a few months.

Probing the Planet

Much of our knowledge of the planet beneath our feet comes from studying seismic waves. Earthquakes are the natural source of seismic shocks. Nuclear explosions are excellent artificial sources. The waves offer the geophysics equivalent of sonograms, with data from seismometers around the world combined to deduce the composition of the material through which the waves propagate.

The unpredictability of earthquakes, in both time and location, forces geologists to rely on serendipity to gather data. Nuclear explosions, fortunately, are rarer today than they were when the weapons were first developed. And the details of their precise timing are typically ensconced in secrecy, to the extent that it's possible to keep such enormously energetic events under wraps. Timing is important for interpreting seismic data, presenting an extra challenge for geologists who hope to use them to study the deep earth on the rare occasion that a nuclear weapon is detonated.

Neutrinos, instead, promise the planetary equivalents of X-ray images and tomographic scans. The medical technology known as positron emission tomography (PET), for example, involves inserting radioactive tracers into a medical subject. The tracers are attached to glucose, water, or other compounds common in the body. Doctors select the compounds based on their likelihood of accumulating in places they need to image, such as the brain or other organs. When an isotope in one of the tracer compounds breaks down through beta decay, it releases a positron that annihilates with a nearby electron to produce a pair of gamma rays. By monitoring the rays, doctors can tell where the radioactive isotope was inside your body. The beta-decay positron is accompanied by a neutrino, but doctors don't currently monitor them in PET scans.

The Earth contains its own supply of radioactive tracers in the form of unstable isotopes of uranium, thorium, and potassium distributed throughout the Earth's crust and interior. As the isotopes break down, they convert to other elements while primarily releasing alpha particles (helium nuclei), gamma rays, and beta particles (electrons or positrons) with their associated neutrinos. The energy and most of the decay products don't penetrate far through the dense planet and are trapped below the ground—all except for the neutrinos. Currently, there's not enough information in limited observations of geoneutrinos, as the ones created inside the Earth are called, to create detailed PET scan–like maps of the planet's interior.

Much more fine-grained images of the planet are possible using neutrinos that impinge on the Earth rather than emanating from it. A nearby supernova, for example, could do the trick. The burst of neutrinos that accompanies a supernova could light up the planet like a camera flash and provide a look at the interior that rivals the decades of geophysical data collected via seismic waves.[5] Of course, we would need to be ready for it. Neutrino detectors around the globe could compare the relative neutrino fluxes and expose the planet's internal features. The more detectors, the better. Ideally, a supernova will occur directly above one detector while others are in the shadow of the planet to varying degrees.

Supernova SN1987A is, so far, the only supernova to occur in the modern era that could have illuminated the Earth's interior. But observatories at the time were at too few locations and captured too few neutrinos to reveal much about the planet's insides. With a little luck, as observatories multiply and experimental methods improve, we should be better prepared to glean geophysical insights when the next nearby supernova detonates.

Some scientists prefer not to wait for a supernova or to make do with the neutrino flash that even a supernova SN1987A-type event can provide. Instead, they've proposed powerful neutrino beam machines that we could use to scan the planet as we see fit. Probably the most ambitious experiment ever to be suggested along these lines was the brainchild of four of the world's leading physicists: Alvaro De Rújula and Georges Charpak of CERN; Sheldon Glashow, who was at Harvard at the time; and Robert R. Wilson of Columbia University. They joined forces to publish a 113-page paper in 1983 describing an undersea particle accelerator to produce a neutrino beam of unprecedented power that could meticulously scan the Earth.[6]

"Earth-scale tomography was certainly what we were aiming for," said Glashow in recalling the proposal. It was a departure from the bulk of his research focus, coming four years after he shared in the Nobel Prize for showing that the electromagnetic and weak fundamental forces are two manifestations of the same thing. "So, [Geotron] was an idea, and we had a lot of fun writing that paper."

It would have needed a beam with record-shattering capabilities. Neutrinos at low energies pass through the planet too easily and provide relatively little information about the interior of the planet. As neutrino energies increase, their interaction cross section also increases, which means they can provide information about the Earth's interior rather than blithely sailing through. The researchers proposed cranking the energy up to a few trillion electron volts to make a beam with a range comparable to the diameter of the Earth. In order to generate such energetic neutrinos, they would need an accelerator ring as much as 24 kilometers across that was capable of accelerating protons to 20 trillion electron volts (20 TeV) before smashing them into a target to produce a beam of particles that would decay into high-energy neutrinos.

The Geotron, as they dubbed the accelerator, was an ambitious concept. The world's highest-energy particle accelerator at the time, the Tevatron in Illinois, which was completed the same year the researchers published their proposal, managed energies only around one TeV. The highest-energy accelerator ever built as of this writing nearly 40 years later, the Large Hadron Collider in Europe, achieves a third of the energy Geotron needed to reach.

Unlike experiments built on stable ground, the submerged machine would have required the magnets that guided the protons around the accelerator loop to be continually monitored and adjusted to correct for the buffeting of ocean currents. A flexible beamline attached to the accelerator—a nose-like projection that the researchers called the Snoot—would guide the high-energy protons to the target where the neutrinos ultimately were to emerge. By precisely manipulating the Snoot with an array of cables, they could sweep the beam through the Earth.

The physicists imagined three lucrative experiments they could perform once the Geotron was deployed. The first, which they dubbed Geological Exploration by Neutrino Induced Underground Sound (GENIUS), could locate oil and gas reserves based on the sounds produced when neutrinos struck atoms along the beam path. The neutrinos would induce showers of particles, which would rapidly heat material in small volumes and cause acoustic pops. They believed the sounds could indicate the type of matter where the interaction took place, and point to potential reserves of oil and natural gas—although there were no data to say for sure whether that was true because neutrinos with high enough energy had not yet been produced in labs to confirm the idea. An array of microphones would monitor the sounds, which the group believed would stand out from the random hum of natural seismic noise.

The second application, Geological Exploration with Muons Induced by Neutrino Interactions (GEMINI), involved truck-mounted muon detectors roaming the area of the Earth where the Geotron beam emerged after passing through the planet. Interactions between the high-energy neutrinos and material near the surface would create bursts of muons, including some traveling perpendicular to the beam. The sideways-moving muons would have a harder time passing through dense materials, which would allow the trucks to identify lucrative ore deposits.

A third application, Geoscan, would rely on muon detectors mounted on oil tankers to monitor muons produced by the neutrino beam as it was scanned through the planet. In a matter of months, the team projected, the scan would have provided a measure of the Earth's density to an accuracy of 1 percent. It would have been a dream come true for geologists pondering the planet's structure.

All of this could be achieved, the four physicists claimed, for $1 billion if they managed to keep the bureaucratic inefficiencies to a minimum. Or if the bureaucrats had their way, the team sniped in their paper, it might cost as much as $2 billion. That would be a range of $3 to $6 billion today. They argued that it was a bargain even at the higher price, both scientifically, because of the potential to learn things about our planet that no other experiment could manage, and financially, by identifying ore and fossil fuel deposits. Ultimately, no one with the access to $1 or $2 billion agreed, and the project never got underway.

"It was an interesting idea," said Glashow while recently reminiscing about the decades-old proposal, "but these days I would not like to be caught responsible for anything encouraging fossil fuel discovery. I think fossil fuels are best left under the ground."

The Geotron was not to be. No human-made machine yet produces the high-energy neutrinos that would be necessary to do the science experiments that De Rújula, Charpak, Glashow, and Wilson envisioned. But researchers have begun to achieve similar goals with natural neutrino sources.

By analyzing data from the IceCube neutrino detector in the Antarctic related to the neutrinos produced by cosmic rays that strike the atmosphere, one group has managed to estimate the mass of the Earth.[7] Although their estimates are less precise than measurements of the planet's mass using conventional geological methods, neutrino data continue to accumulate.

Some researchers have suggested that comparing the results from the two types of experiments, once the neutrino measurements are precise enough, could give us insights into dark matter. After all, gravitational measurements of the planet's mass should include the contribution due to dark matter because the one force that affects dark matter is gravity, as far as we know.

If neutrinos interact only with matter, using them to estimate the Earth's mass should reflect the contribution due exclusively to normal, non-dark matter. The difference between the two mass measurements could tell us how much dark matter is in the Earth. Current estimates based on astronomical observations suggest that about four parts in 10 million trillion is dark matter.[8] It will require very precise measurements indeed to see any sign of dark matter by comparing gravitational and neutrino measurements of the Earth's mass.

A few of the proposed Geotron applications may now be accessible without the need to generate high-energy neutrinos, thanks to the confirmation of neutrino oscillations. But it wasn't until well after the Geotron proposal that it was confirmed that the neutrinos shift from one variety to another. What's more, we now know that the oscillations change when neutrinos interact with matter, and that the changes depend on the density of the matter the neutrinos pass through.

Instead of listening for acoustic signals that indicate the types of material along a high-energy neutrino beam path, counting neutrino varieties in a beam fired through the Earth would reflect varying material densities. Although the technique would be less likely to give specific information about the material the neutrinos encounter, they could suggest promising regions to drill for fossil fuels.[9]

Hunting for oil deposits by monitoring neutrino oscillations would be much cheaper than the GENIUS approach because the ideal neutrino energy range is in the hundreds of millions of electron volts, rather than the trillions of electron volts that the Geotron was designed to produce. However, as the emphasis on addressing climate change reduces the motivations for uncovering new, carbon-based fuel deposits, the financial motivations have dwindled as well.

On the other hand, interest in understanding the structure of the Earth persists. Probing the planet by way of neutrino oscillations could complement seismic and gravitational measurements by providing a clearer, more direct measure of the planet's density.[10]

Earth Engine Diagnostics

Compared to the other rocky planets in our solar system, the Earth is unusually active. We live on continents that drift about on a layer of hot rock; volcanoes erupt and earthquakes strike unpredictably; and we benefit from a comparatively intense magnetic field that protects us from deadly radiation from space. All of this is due to the heat and flow of material deep underground.

When scientists first pondered the dynamic interior of Earth, the source of the energy powering the activity was a mystery. One of the best guesses came about in the mid-1800s when Lord Kelvin proposed that the heat in the planet's interior stemmed from the gravitational energy of the matter that initially came together to form the Earth. Kelvin estimated that the planet started as a molten ball at a temperature of about 3900°C. Over time, the Earth cooled as heat flowed out. It was a conceptually simple matter of measuring the temperature of the Earth at various depths underground to figure out how much heat remained and how long it had been since the planet formed and first began to cool. The estimate Kelvin derived for the age of the Earth, based on those measurements, was tens to hundreds of millions of years. His calculations were far short of the numbers even geologists at the time put forth, and nowhere near the 4.5 billion years we now know has passed since the Earth formed from the dust in orbit around the Sun.

In large part, Kelvin's estimate of the planet's age was flawed due to his overly simplified model of the Earth's structure. It is not, as he supposed, a static solid ball, with heat simply diffusing outward. Instead, as the drift of the continents, the planetary magnetic field, and other effects show, the interior is in constant motion through the convection in the mantle. The complexities of the Earth's interior allow a much greater store of heat than the simple measurements of temperature with depth seem to indicate, at least for the relatively shallow measurements available at the time. Taking that into account, as one of Kelvin's contemporaries did,[11] led to significantly higher calculations of the planet's age, some much greater than current estimates.

The debate over the Earth's age was at its peak around 1896 when Henri Becquerel discovered radioactivity. Marie Curie's discovery a few years later of radioactive elements including polonium and radium radically altered the debate over the age and structure of the Earth. For one thing, it made it possible to date rocks by measuring the ratios of radioactive elements and the products they become when they decay. Knowing the rate of radioactive decay made it possible to calculate ages of rocks to billions of years, offering a way to directly determine the age of the Earth and to definitively show that Kelvin's estimates of 10–20 million years were far too brief. In addition, the discovery that radioactive material can produce substantial amounts of energy made it clear that the assumption that the planet is steadily cooling is not correct; radioactive material in the Earth's crust and mantle contribute significantly to the heat.

Clearer pictures of the planet's structure through the study of seismic waves and firmer measurements of the Earth's age, settling on about 4.5 billion years, have made it apparent that there is unquestionably a source of energy keeping the interior hot and that the source must be the decay of radioactive elements. Just how much radiation contributes to the heat is still a matter of debate, largely because there has not been a good way to directly measure the radioactivity of the planetary interior.

The complexity of the Earth's structure and the challenges of modeling it lead to estimates of the total planetary heat emission that range widely, from 14 to 47 terawatts (trillion watts), which is on par with the total worldwide human power consumption of 15 terawatts. About half of the Earth's power output is likely due to internal radiation. The rest is the result of the residual heat of the planet's formation, which means, in retrospect, that Kelvin wasn't half wrong.[12]

The heat that the decays produce is distributed throughout the mantle over time, hiding its origins. The particles that are the decay products of the radioactivity, as well, don't make it very far through the dense mantle. The deepest boreholes for retrieving rock samples that might give some insight into the distribution of radioactive elements have reached a little over 12 kilometers down. At a few tenths of a percent of the radius of the planet, they barely scratch the surface, penetrating only a third of the way through the crust and coming nowhere near the flowing mantle. The only direct way to monitor radioactive materials deep underground in real time is by capturing the geoneutrinos they emit.

About 20 percent of the planet's energy radiates away as geoneutrinos. They come mostly from radioactive isotopes of uranium, thorium, and potassium. The elements tend to bind well with silicate materials that are the main components of the Earth's mantle and crust, rather than the metal that makes up the planet's dense central core. Although the mantle is solid rock, it flows slowly over long time periods. The shifting of tectonic plates, volcano activity, earthquakes, and the Earth's magnetic field are all effects stemming from convection in the mantle.

Although potassium isotopes are responsible for a significant fraction of the energy and geoneutrinos emitted from the Earth, their energy is too low to register in contemporary detectors. Neutrinos from uranium and thorium decays, however, are energetic enough to be detectable.

The numbers of geoneutrinos detected annually are low, in the tens to hundreds per year. Experiments on the horizon will increase the detection rates dramatically. Even at current levels, however, there are enough geoneutrino detections to suggest that the heat energy the Earth emits is at the higher end of the estimates, around 38 terawatts, and that natural radiation from unstable uranium, thorium, and potassium isotopes likely accounts for about half of that.

Neutrinos emanating from the Earth, or rather the lack of them, have put limits on a proposed solution to one of the lasting mysteries in geophysics: the origin of the planet's magnetic field. Electrical currents in the Earth's metal core and the flowing, conductive material in the mantle create the planetary magnetic field. The energy that drives the electrical dynamo is generally ascribed to the heat flowing out from the metal core and the radioactive elements in the mantle. An alternative explanation, primarily developed and championed by geophysicist J. Marvin Herndon, is that the engine of the magnetic field is driven by a natural nuclear reactor at the center of the Earth.

Herndon has supposed that a natural reactor could account for the magnetic field, as well as volcanic activity and continental drift. It would have to be a three- to six-trillion-watt reactor, roughly a thousand times the power of a large commercial reactor, and eight kilometers or so in diameter. A reactor in the Earth's core could also potentially account for the abundant supply of helium underground, which would be a product of the natural nuclear reactor process.[13]

Unfortunately for visions of a natural reactor powering the planet's dynamic interior, the neutrinos that would come from a reactor of the size Herndon proposed don't turn up in detectors. The Borexino collaboration's research, published in 2020, finds with 95 percent confidence that there is no nuclear reactor at the center of the Earth any more powerful than 2.4 terawatts.

Assuming there is a natural reactor somewhere, but not at the planet's center, it's possible that Borexino is simply too poorly positioned to tell and that a natural reactor in the three to six terawatt range exists somewhere else underground. Increasing numbers of detectors that will be coming online soon, including some detectors more sensitive to low-energy geoneutrinos, should give us a better indication of whether or not there is a reactor powering the Earth's dynamo. We can't say for sure at the moment, but the outlook has become increasingly unlikely as we become more adept at observing neutrinos originating from inside the planet.

Spin-Offs

Neutrino detectors often are sensitive to things other than the little neutral ones. A variety of hypothetical dark matter candidates have been proposed to solve some outstanding problems in physics. If they exist, they may turn up in neutrino detectors. Currently, Borexino is among the most prominent experiments that are potentially sensitive to some proposed types of dark matter.

IMB, which was one of the three detectors to see neutrinos from supernova SN1987A, was built initially to look for signs that protons can spontaneously break apart. Proton decay would potentially help explain why the universe is primarily made of matter rather than an equal amount of antimatter and matter. More recently, the neutrino detector Super-Kamiokande has become the most sensitive measure of proton decay. Theory suggests that when the proton decays it will produce a positron and a pion. The pion would then rapidly decay into a distinctive pair of photons, which will appear in Super-K's photodetectors.

In Super-K's 50,000-ton tank of ultrapure water, there have been no signs of proton decay over a decade of observations. Considering that each water molecule includes 10 protons (one in each hydrogen atom and eight in the oxygen atom in a molecule of H_2O) and there are 30 billion trillion water molecules in a gram, and a million grams in a metric ton, the fact that not a single proton decay has turned up in Super-K means that protons live a very long time before they break apart, if they do at all. The result is consistent with some theories that predict the typical lifetime of protons. Hyper-Kamiokande, a successor to Super-K due to come online in 2027, should be sensitive enough to proton decay to start testing many theoretical predictions.

Some neutrino physics spin-offs, however, have little to do with particle physics at all. The first detector that neutrino discoverers Cowan and Reines built failed to find neutrinos from a nuclear reactor, but it had important applications for the new weapons program in Los Alamos. Although they saw tantalizing hints of neutrinos in their 1953 experiments, the signals were buried in spurious noise due to gamma rays. They would need to redesign their detector for their next attempt. But when they returned to their labs in Los Alamos, they realized that they could repurpose the neutrino experiment to take advantage of its sensitivity to gamma rays to fill an important health physics need in fledgling nuclear industries and weapons programs.

Radioactive materials create inherent risks for workers. The dangers often come in the form of direct exposure to X-ray, gamma ray, or particle radiation. In those cases, radiation monitors, such as dosimeter badges and Geiger counters, offer estimates of radiation exposure and provide warnings so that workers can be moved to safe distances or to areas shielded from radiation.

Sometimes, the materials are in the form of radiation-emitting gases or liquids that workers inhale or absorb, leading to internal, potentially persistent bodily contamination. In order to track and treat internal contamination, it's vital to have a way to measure the radiation from a person's entire body.

Due to the size and design of Herr Auge, the first detector Cowan and Reines built, they realized that it could fit an adult human inside. They inserted a smaller cylinder in the middle of the detector that could accommodate an adult subject, provided they were willing to crouch and suffer being enclosed in a covering layer of lead bricks. The space between the inner and outer cylinders was filled with the liquid scintillator that puts out flashes of light in response to radiation. The flashes in turn would register in the light sensors mounted in the outer wall of the tube, just as they were intended in neutrino experiments.

Instead of counting neutrinos, though, the detector registered gamma rays from radioactive contaminants in the subject's body. The gamma rays that were a source of noise drowning out data in Project Poltergeist became the important diagnostic signal for their whole-body radiation counter.[14]

Eight staffers at the Los Alamos Science Laboratory,[15] identified only by their initials in the published research, took turns crawling into the cramped detector to have their whole-body radiation levels counted. Two of the subjects who handled radioactive materials including radium and thorium as part of their duties were tested after coming straight from work. Their total radiation counts exceeded other subjects by factors of 10 to 25. When the clothes of one of the workers was tested separately, it became clear that the garments were responsible for the bulk of the contamination.

(a)

(b)

A 1967 U.S. Atomic Energy Commission pamphlet on whole-body radiation counters includes images of Los Alamos Scientific Laboratory staff enduring experiments to see if the Herr Auge neutrino detector could serve as an effective tool for measuring radioactive emissions from humans. (a) Dr. Frederick Reines (left) and Dr. Clyde L. Cowan (right), co-discoverers of the neutrino, lower a fellow worker into the first "whole body counter," the scintillation assembly used in their experiment. (b) Dr. Wright Langham, inside the counter, peers from the opening. *Source:* Frederick W. Lengemann and John H. Woodburn, *Whole Body Counters* (United States Atomic Energy Commission, Division of Technical Information, 1964; rev. 1967).

A ninth subject, and the only one named in the study, was physicist Wright Langham, who would later become one of the leaders in the Biomedical Research Division at the lab. Langham carried a sample of radium into the detector, clutched it to his stomach, and shielded it with his body. Although it was not inside him, the radiation would have to pass through his body to make it to the detectors. It confirmed that detectable radiation would turn up in the repurposed Herr Auge neutrino experiment after passing through a human.

Even without simulated or actual contamination, the detector could measure subjects' potassium levels to within 14 percent by monitoring emissions from the radioactive isotopes of potassium that naturally make up a small fraction of the element in the human body. For all of the subjects in their study, except one with a history of working with radioactive materials, Cowan and Reines concluded that the radiation they were measuring was the result of potassium. Considering that a potassium deficiency (hypokalemia) can lead to serious health problems, Cowan and Reines proposed that their detector could provide early warning of a potassium-related health problem, which might be the case if a patient had an unusually low level of whole-body radiation.

Injections of solutions containing radium into beagles, which were then anesthetized and placed in the detector, provided a clearer indication that whole-body radiation counters would be effective diagnostics for exposure to radioactive materials. It wasn't long before the Los Alamos labs produced a scaled-up, medical-grade version of the counter. It allowed patients to stretch out at full length as they were slid into a gamma ray counter. Although the medical system was based on the neutrino detector technology Cowan and Reines pioneered, it would take another three years after using it to check whole-body radiation before they put the same basic design to work to actually find neutrinos.

More recently, optical noise in the ANTARES neutrino experiment is offering insight into deep-sea life. The experiment's primary purpose is to search for high-energy neutrinos coming from distant cosmic sources. It's a detector array that consists of 12 vertical strings of 75 photodetectors moored to the floor of the Mediterranean Sea near the coast of France. As high-energy neutrinos pass through the seawater, they occasionally create muons that initially travel faster than the speed of light in water. A flash of Cherenkov light reveals the neutrino's interaction and direction. They are not, by any means, the only source of light in the ocean. Many deep-sea organisms produce light as well, which the photodetectors pick up as easily as they do Cherenkov light.

The light coming from natural, living sources differs depending on the creatures that cause them. The ANTARES team's biology colleague, Christian Tamburini, has speculated that longer periods of light are due to populations of bacteria, while sea jellies, shrimp, and luminous fish make briefer flashes.[16] In order to distinguish between animals and neutrinos, the researchers installed cameras that capture images of light-emitting creatures near the photomultiplier tubes in the neutrino detector array.

Sensors that monitor salinity, oxygen, and sea currents are important for the neutrino scientists who need to understand the environment around their detectors. Along with the cameras, they also provide crucial information for oceanographers and marine biologists.

Microphones on the ANTARES detector strings that make up AMADEUS (ANTARES Modules for Acoustic DEtection Under the Sea) are designed primarily to detect the sounds that high-energy neutrinos produce underwater when they deposit energy that leads to audible pops (like the ones the Geotron would have generated in oil deposits). The microphones pick up other noises as well, including the vocalizations of sperm whales. Most surveys of whale sounds rely on short-term, battery-powered monitoring systems. Because ANTARES and AMADEUS are powered through dedicated cables, the data are available full-time. The long-term observations have allowed marine biologists to confirm that sperm whales are present year-round in the Ligurian Sea portion of the Mediterranean between Corsica and the coast of Italy. The scientists were surprised to find, thanks to the continual monitoring the neutrino detector provides, that sperm whale foraging activity appears to be relatively unaffected by shipping traffic in the region.[17]

More Promise than Purpose, for Now

Six decades after the discovery of the neutrino, and nearly a century after its initial hypothesis, we are just beginning to see pragmatic, technical applications of the ghostly particle materialize. Monitoring nuclear reactors and probing the Earth are among the first areas where the neutrino's unique properties might be exploited to address technological problems.

Considering that some of the leading physicists of the last century would have bet that we'd never be able to detect neutrinos at all, and now we have a least a few applications that are on the cusp of fruition, it might be unwise to bet against technological solutions and applications in the long run, even where the elusive neutrino is involved.

15

On the Fringes and over the Edge

Let's think the unthinkable, let's do the undoable. Let us prepare to grapple with the ineffable itself, and see if we may not eff it after all.

—Douglas Adams, *Dirk Gently's Holistic Detective Agency*

Neutrinos are mysterious today. They were even more so in the years immediately following their initial discovery. That may in part explain one of the first attempts to apply them to at least one, dubious technological application: telepathic communication.

In the mid-1960s, according to a report commissioned by the U.S. Defense Intelligence Agency,[1] Soviet paranormal researcher Ye. Parnov proposed a theory that relied on neutrinos as carriers in what he called a telepathic field. Parnov postulated that the field linked all humans together, allowing us to communicate directly from mind to mind and to see the future.

Parnov's neutrino telepathy theory was just one of several projects that were the subjects of Soviet and Czechoslovakian paranormal research at the time. To put it in perspective, other notable efforts in the paranormal science community behind the Iron

Curtain included a form of instantaneous communication based on murdering bunnies in a submarine at sea, methods to telepathically attack personnel in U.S. and allied nuclear missile silos, and devices known as psychotronic generators that were supposedly capable of such impressive feats as putting snails to sleep and killing flies at very short range. Neither the neutrino telepathy theory nor any other of these projects has gained any traction in the last half century.[2]

Although we don't yet know all there is to know about neutrinos, it's clear that they don't have a role in telepathy, extrasensory perception, or telekinesis. After all, your brain would need to include an accelerator or large supplies of radioactive isotopes to produce neutrinos in any significant quantity. And your head would have to be at least the size of a small truck to detect them. Nevertheless, esoteric and mysterious neutrinos potentially promise exotic applications. Some may, at first, seem as bizarre as neutrino-based parapsychology, which is a testament to just how odd and interesting neutrinos are, even without invoking pseudoscience.

Ultra-High-Speed Global Communications

The shortest distance between two points is a straight line, and the universal speed limit is the speed of light. If you want to send a signal from one point to another, photons are your best bet—unless something gets in the way, like, for example, the planet.

The circumference of the Earth is about 40,000 kilometers. If you were to send a signal to the opposite side at the speed of light, a message traveling just above the planet's surface would arrive in about 66 milliseconds. In order to get a light beam from you to your distant destination, though, you would need to guide it around the sphere of the Earth. One way to do that might be an arrangement of mirrors to bounce the light from point to point relatively close to the surface. That would take a bit longer, but still allow you to send a signal halfway around the world in a hundred milliseconds or so.

In many cases, we communicate by way of wires or optical fibers, rather than relying on mirrors, when we need to get signals around obstacles and curves. Light travels slower in optical fibers than in air by about 30 percent or so. A message passed through a fiber-optic cable strung from Hong Kong to La Quiaca, Argentina, on the opposite side of the world would take 85 milliseconds.

In practice, though, global telecommunications systems involve a mix of wires, fibers, and satellites that relay signals from point to point high above the Earth's surface. There are delays as signals pass between satellites and through optical and electronic switches, which means it can take about a quarter of a second for a message to make its way from Australia to the United States.

If, instead, you could send a message directly through the Earth at the speed of light, it would arrive in a scant 43 milliseconds. Short of drilling a hole through the center of the planet, the only methods with potential to send messages that quickly are by way of neutrinos or gravitational waves.

We can detect gravitational waves coming from colliding black holes and other massive objects, but we're unlikely to be able to create gravitational waves powerful enough to send useful signals any time soon. That leaves neutrinos.

Point-to-point neutrino communications that can penetrate directly through any obstacle, or planet, that happens to stand in the way were proposed in published science journal articles as far back as 1977.[3] Even decades later, modern neutrino sources and detectors wouldn't work well for sending signals from one side of the globe to the other—the sources are too meager and the detectors too low in sensitivity. That didn't stop researchers from taking the first steps in neutrino messaging. In 2010, North Carolina State University engineering professor Daniel Stancil led a team that co-opted the Fermilab neutrino beam and a detector located a little over a kilometer away to send a brief, slow, message. The system is typically dedicated to studying neutrino oscillations, but for a few hours, the researchers flicked the beam on and off in a pattern that encoded the word "neutrino."

Thirty-three years after it first appeared in the formal science literature, and half a century since it was first bandied about informally, someone had at last sent a message via neutrinos. It was not a widely hailed achievement. "We received a lot of pushback about doing the experiment and, after doing it, getting it published," says Stancil[4] in describing the effort to make the first ever transmission by what he called neutrino radio. "There were those in the physics and engineering communities that took the position that there was no new information about neutrinos nor communications theory here. However, from an engineering perspective, building and demonstrating a concept with existing technology is a critical step in identifying challenges and determining where advances are needed. So, I absolutely believe it was an important thing to do, and having demonstrated the concept, I believe it has encouraged others to consider more carefully what may be possible."

The neutrinos that carried the message passed through 240 meters of rock, primarily consisting of shale. The single word message was repeated multiple times over the course of 142 minutes, with each instance taking a little over a minute to send.[5]

Communications are generally measured in hertz, which reflects how many bits can be transmitted per second. Modern high-speed systems can send information at a billion bits per second. Human speech typically transmits messages at a rate that translates to roughly 39 bits per second.[6] The Fermilab experiment achieved a blistering data rate of one bit every 10 seconds, 400 times slower than human speech. In effect, you could have more quickly sent the information over a kilometer by shouting, or on horseback for that matter, than the Fermilab experiment managed. The key difference, of course, is they did it through solid earth—try that on a horse.

Although successfully transmitting the neutrino beam equivalent of Alexander Graham Bell's "Mr. Watson, come here; I want you," the Fermilab demonstration made it clear that the approach is in no way ready for practical communications. For one thing, the source of neutrinos was enormous: protons accelerated through a series of steps, ending in an accelerator ring more than three kilometers in circumference, and then smashed into a carbon target, which in turn produced a beam of particles that decayed into neutrinos and other particles. The detector too was huge. The Main Injector Experiment for ν-A (MINERvA) detector that received the message consists of five tons of scintillator strips and light-detecting photomultiplier tubes.

You won't be placing point-to-point neutrino calls directly through the globe any time soon. Saving the quarter-second delay for a chat between Tokyo and New York is unlikely to be worth building dedicated neutrino detectors and particle accelerator transmitters. Financiers, on the other hand, have pursued informational advantages for as long as trade has existed. At one time, access to the fastest carrier pigeon, carriage, or boat was enough for an investor to stay ahead of the competition. The invention of the telegraph, radio, and microwave antenna, among numerous other technological advances over the years, contributed to increasingly speedy communications between financial centers.

In a paper published to the journal *Financial Review* in 2014,[7] physicists Gregory Laughlin and Anthony Aguirre of the University of California, Santa Cruz, and law professor Joseph Grundfest of Stanford estimated that improvements to communication links between the Chicago and New York financial markets intended to speed trading and reduce information latency were running at $500 million every five years.

"In the distant future," they wrote in concluding their paper, "we can speculate that exotic technologies such as neutrino or even WIMP, axion or gravity wave communications could be employed to communicate financial tick data directly through the Earth." Perhaps tens of billions of dollars to build a neutrino system that can reach any point on the planet a few, critical tenths of a second ahead of any other known technology might be worth the investment to some investors and may someday make global neutrino communications lucrative.

There is at least one sober application of neutrino communications that has few, if any, decent alternatives: direct, high-speed communications with submarines. Extremely low frequency (ELF) radio wave signals generated with enormous antennas can transmit limited instructions, potentially commanding the subs to send a conventional antenna-equipped buoy to the surface, or perhaps to head to an underwater "phone booth" linked to the land via fiber-optic cable. And, of course, subs could communicate via acoustic signals as some marine creatures do, at the risk of giving away their position to anyone else who might be listening. Neutrinos, in theory, could provide an alternative that avoids some of these short-comings, although laden with a different set of challenges.

The bit rate of Stancil's neutrino radio experiment at Fermilab was excruciatingly slow, even by ELF standards, but that was chiefly due to the pulse rate of the neutrino beam. The accelerator that created the beam provided 8.1 microsecond pulses every 2.2 seconds, with the repetition rate limited by the time it takes to accelerate the protons that create the neutrino beam.

The highest-speed communications achievable with neutrino beams and detectors like those at Fermilab, assuming they were reconfigured as transmitters and receivers for communications rather than a research device, could be between one and 100 bits per second. That would give one-bit-per second ELF some tough competition, though still limit messages to brief instructions.

The key difference between undersea electromagnetic radio communication and neutrino radio is that the impediments to the first are the result of fundamental scientific limits. Electromagnetic signals don't penetrate water well. You'd have to build very large, powerful transmitters to send excruciatingly slow signals. Neutrino radio limitations are a matter of technology. And we can eventually overcome technological impediments, at least in principle. If the navy ever wants to send more than brief, slow messages to submarines actively prowling the deep oceans, neutrinos are the only carrier that can even theoretically make it possible.

The Search for Aliens in Space

Intelligent life should be common in the universe, at least according to people who take to heart some versions of an equation that astronomer and astrophysicist Frank Drake proposed in 1961. Drake's equation combines things we know well about the universe (the rate of star formation) with things we are learning more about (the fraction of stars that have habitable planets near them) and things we know very little about at all (the likelihood that life will develop on a habitable planet, and the time it typically takes for intelligent life to evolve, produce detectable signs of their existence, and send signals into space). Most of the best guesses that go into the equation imply there is almost certainly intelligent life out there somewhere.

Drake's equation has been the subject of discussions, arguments, refinements, and proposed corrections over the decades. Perhaps the greatest challenge the equation has faced is a simple question Enrico Fermi posed: "Where is everyone?" If Drake's equation is correct, the universe should be teeming with intelligent life.[8] And if that's true, we should have seen evidence by now.

It's a big question, and covers a complex array of factors, including those in Drake's equation and more. One open issue is just how a distant alien might choose to send us a message, if they were so inclined. At first glance, a simple electromagnetic signal, in the form of radio, visible light, X-ray, or gamma ray photons might come to mind. Unfortunately, while the space between stars and galaxies has much less material in it than we are used to in our atmosphere on Earth, space is a great deal murkier than you might imagine. In the course of its travels from a distant planet a beam of photons will encounter dust and plasma that would likely distort and obscure signals.

Another possibility that some researchers have proposed is communication via gravitational waves. Since 2016 we've had the ability to measure gravitational waves with a pair of enormous experiments known collectively as the Laser Interferometer Gravitational Wave Observatory (LIGO). Each of the LIGO labs, one located in Hanford, Washington, and the other in Livingston, Louisiana, consists of L-shaped arrangements of laser beams bouncing back and forth in tunnels four kilometers long in order to measure ripples in space. The ripples indicate the passing of gravitational waves. Black holes and neutron stars colliding billions of light years away created the gravitational waves that LIGO has detected so far.

Theoretically, an advanced alien could send a message encoded in gravitational waves. The waves would pass easily through the intervening, dust-filled space and we could observe them with LIGO or, more likely, a much more sensitive, subsequent-generation gravitational wave observatory. To produce signals we could measure, the alien would need to generate them with transmitters built of black holes or other super massive components—a difficult but perhaps not impossible task for a sufficiently advanced alien.[9]

Neutrinos provide an alternative for communicating over galactic and cosmic distances that combines some of the benefits and challenges of both photons and gravitational waves. Like gravitational waves, they can propagate over enormous distances without being significantly scattered or absorbed by interstellar dust or, for that matter, asteroids, planets, or even stars.[10] Like photons of light and other electromagnetic radiation, aliens could produce neutrinos in beams that they can aim at promising regions of space where a message might be most likely to be received. This sort of targeting is much more efficient than broadcasting widely, if you have some idea of where to point it.

"I think it's not crazy that some very advanced civilization would use neutrino beams to communicate," says Anthony Zee[11] of the Kavli Institute for Theoretical Physics in Santa Barbara, California. "It has obvious advantages. It travels at nearly the speed of light, it's not attenuated, it goes through everything. I personally think that instead of spending zillions and zillions on these huge antennas" that are the primary tools in searching for signs of extraterrestrial intelligence, "these guys should fund some neutrino detectors. It wouldn't even have to be a big detector; it could just collect data and see if you hear anything."

As is true for light and gravitational waves, neutrinos travel at finite speeds, which means that anyone using them to send a directed signal will have to think ahead a bit. If the star system you're messaging is a million light years away, then you'll have to point your neutrino beam to the spot where your target will be in a million years, because neutrinos travel at speeds comparable to the speed of light. Based on the distribution of potentially habitable planets, even inside a single galaxy, two-way communication will suffer some serious delays.

Among the primary arguments against aliens using neutrinos to communicate—even if you assume they have the technology do it—is that, all in all, it's simply much easier and cheaper (in terms of the energy involved) to send signals with photons than with neutrinos or gravitational waves.[12] A large percentage of photons will be scattered or absorbed, but an alien could always compensate by turning up the power in their transmitters, with a lot less effort than creating neutrino beams.

One way to turn photon signal power way up, according to Zee and neutrino experimentalist John Learned, might still rely on neutrinos.

In a paper titled "The Cepheid Galactic Internet," they describe a system that could theoretically change the blinking of Cepheid variable stars that have naturally fluctuating light outputs.[13]

"It turns out the reason they're blinking is they're on the verge of instability," says Zee, which means that a small nudge could modify the variation in a Cepheid star's fluctuations. "The idea is to shoot a neutrino beam off center into the star and trigger some sort of instability." It's the particle physics equivalent to using the noise from a gunshot to start an avalanche on an unstable snowfield. "We call it tickling the star, tickling the Cepheid variable," says Zee. "The simple idea is that an advanced civilization would be able to capture the output of the star and somehow store it and dump it into a neutrino beam as a pulse . . . you can do some estimates, and it's not completely crazy. If a planet orbiting a star could collect even one percent of the energy output per unit time, you can of course multiply that by storing it . . . and then pulse it back into the star." According to Zee, "It would act like a universal beacon."

"We propose that these (and other regularly variable types of stars) be searched for signs of phase modulation (in the regime of short pulse duration) and patterns, which could be indicative of intentional signaling," the physicists write in their paper's abstract, on the chance that some distant intelligence is already uploading information to the Cepheid Galactic Internet. They conclude the paper by writing, "It may be a long shot, but should it be correct, the payoff would be immeasurable for humanity. The beauty of this suggestion seems to be simply that the data already exists, and we need only look at the data in a new way."

Catching Cosmic Engineers at Work

Even if aliens don't choose to communicate via neutrinos or neutrino-modulated stars, the particles could still reveal their presence in the universe as a side effect of alien activities. It seems reasonable to guess that intelligent life, wherever it is, would have a similar thirst for knowledge that humans have. It's conceivable that advanced aliens might rely on particle accelerators to conduct research, as physicists around our world currently do.

We already know that accelerators powerful enough to explore the ultimate limits of nature, at a tiny scale called the Planck limit, would require accelerators far too big to fit on a planet. The smaller scale you want to study with an accelerator, the higher the energy you need to accelerate particles, and the larger the accelerator must be.

In a paper published on the ArXiv in 2015,[14] Brian Lacki, then at the Institute for Advanced Study in Princeton, New Jersey, estimates that an accelerator that could explore nature at the Planck limit would be as large as a typical solar system. If we could find hints of one running, it would be a clear sign of super intelligent beings, which Lacki terms "cosmic engineers," somewhere in the universe. High-energy neutrinos could provide that sign.

Smashing together matter, whether in a terrestrial accelerator or in one built by cosmic engineers, creates enormous showers of particles. By studying the showers, scientists can gain insights into the structure of subatomic particles, recreate conditions in the early universe, and search for new particles that may supply clues to science beyond the models we now know about.

In a Planck accelerator, the showers would include neutrinos with energies far greater than those that could be produced in even the most violent astronomical events, potentially providing a beacon of the sentience responsible for creating the accelerator. The highest-energy neutrinos so far detected on Earth reach thousands of trillions of electron volts, while those from a Planck accelerator would be a billion times more energetic still.

Any beings capable of building a Planck accelerator could probably find a way to capture even neutrinos if they wanted, rather than broadcasting to the universe that their experiments were underway. Keeping the high-energy neutrinos under wraps would be tricky. One way to do it would be to pile up something dense, like lead, at the end of the accelerator to catch the beam before the neutrinos sail into space. It would have to be a big piece of lead to do it, probably about as large as the accelerator itself. The downside to catching the beam is that it would deprive alien researchers of potentially interesting science that could result from the super-high-energy neutrinos crashing into other matter in the universe. They would have to be pretty committed to staying invisible to try to keep the neutrinos out of sight.

It's also unlikely that aliens would keep such an accelerator running beyond the time it took to gather data from an experiment. On Earth, we mothball and decommission experiments once they've supplied all the results we can hope for. It's hard to imagine what goes through an alien's mind, but it seems likely they would do the same. Once their experiments are completed, they would probably turn them off and go in search of other things to study. If that's so, neutrinos coming from solar system–sized accelerators would be fleeting, as alien beings around the universe built them, ran experiments, and shut them down.

Still, if we were to discover such high-energy neutrinos, they would suggest either that there are advanced accelerator scientists at work somewhere in the universe or that there are some exciting natural things going on in the cosmos that are beyond the physics we currently understand. Either way, there's a case to be made to design detectors on our planet to look for the ultra-high-energy particles.

Neutrinos could also reveal alien engineered structures known as Dyson spheres. Visionary physicist Freeman Dyson proposed that as civilizations advance, their need for energy would increase beyond the relatively meager proportion of energy coming from a star that falls on a planet comparable to Earth. Most of the energy from our sun streams out into space. One solution, Dyson suggested, could be to enclose a star in a sphere that collects all the available energy a star produces.[15]

Although a Dyson sphere would prevent us from seeing the light and much of the rest of the radiation coming directly from the star inside, that doesn't render the sphere itself invisible, at least in the simplest implementation of such a structure. The waste heat that's left over after the residents of the sphere have extracted useful energy from their entrapped star has to go somewhere. It would warm the outer sphere, which in turn would glow as a result of the radiation that is emitted from any warm object (this is why we humans turn up clearly in infrared images). If the inhabitants of a system inside a Dyson sphere prefer to keep a low profile, they could pipe the heat away from the bulk of the sphere and send it in a direction where they're fairly certain there is no one to observe it.

Neutrinos, however, would give them away. Even a Dyson sphere with a wall thickness comparable to the radius of our solar system would trap only a tiny fraction of the neutrinos coming from a star like our sun. If an alien tried to build such a thing from material comparable in density to the earth, the total mass would be a trillion times that of the sun, in which case it would collapse under its own gravity and become a black hole, crushing everything inside it. Dyson spheres generally are fanciful ideas and would be hard to build at all, while making one big enough to ensure that neutrinos don't reveal the star inside is simply impossible.

We don't have neutrino telescopes capable of seeing an individual Dyson sphere at any great distance, but if one were to try to surreptitiously glide past our solar system, the neutrino flux would potentially turn up even in existing detectors like Super-Kamiokande. Of course, the gravitational interaction would throw our solar system into disarray as well, but that's a topic for another book.

Improbable, but Maybe not Impossible

Parapsychology is not an area where neutrinos seem to be relevant or useful. Many other potential uses may appear almost as bizarre, even if they aren't prohibited by physics. The odds are slim that we will communicate with aliens via neutrino beams. And transmitting financial data through the planet is not yet on the horizon. Neutrino-based technology, however, may ultimately be the best and, in some cases, the only way to manage such feats.

Will neutrino applications like these ever be possible or justifiable? Maybe. For now, it's probably more the stuff of science fiction. Bear in mind, though, the same could once have been said of airplanes, lasers, travel to the moon, and Pauli's desperately proposed neutrino.

16

More to Come for Little Neutral One

Each new discovery broadens our knowledge and deepens our understanding of the physical universe: but at times these advances raise new and even more fundamental questions than those which they answer.

—F. Reines and C. L. Cowan[1]

Our revels now are ended, but the same is emphatically not true for the physics of neutrinos.

"It is indeed an exciting time for neutrinos," says Kate Scholberg, whose career as a neutrino physicist has spanned the study of supernova neutrinos, the discovery of neutrino oscillations, and the pursuit of coherent neutrino scattering, among other topics. "Over the last few decades, the basic three-flavor picture has emerged from the gloom, but the details are still not clear, and it will probably take still a few more decades to clarify the picture. And the details of the eventual picture might well be weirder than we imagine—there are many possible places where new physics could be hiding in the neutrino sector."

One goal is to fill in the blanks on the fundamental properties of neutrinos that still are not understood: Where do neutrino masses come from, and how big are they? What is the hierarchy of active neutrino masses? Are neutrinos their own antiparticles, or not? Do sterile neutrinos exist? What is the explanation of oscillation anomalies in short-range experiments?

Another goal is to probe the farther reaches of physics.

"Neutrinos are everywhere—and because they're the Houdinis of the particle world, they can escape from strange places with lots of information," says Diana Parno of the KATRIN collaboration. Neutrinos are uniquely positioned to tell us more about supernovas (assuming a star in or near our galaxy is considerate enough to explode), about the early universe, and perhaps about dark matter as well. In addition, there is the exciting prospect of multi-messenger astronomy, the ability to study exotic phenomena simultaneously with photons, neutrinos, and gravitational waves.

In all these endeavors, experimentalists face the daunting challenge of overcoming the extreme weakness of neutrino interactions. This was clear from the earliest days: It took 25 years from Pauli's original suggestion before Cowan and Reines were able to detect them, and Ray Davis fought a decades-long battle in the Homestake mine to determine how many neutrinos are coming from the sun. Their successors today are contending with minuscule event rates and potentially overwhelming backgrounds as they search for neutrinoless double beta decay and try to pin down neutrino masses. Detection of the cosmic neutrino background is still a dream, with no clear path yet to realize it.

Despite the intense activity, new knowledge accrues slowly. Often it takes multiple experiments to obtain definitive results. After the Davis experiment, it took SNO to confirm the oscillation hypothesis, and KamLAND to pin down the oscillation parameters. After the LSND anomaly, researchers turned to MiniBooNE to resolve the issue, but together the two experiments leave major questions that will require further investigation. As you might imagine, the first claim to observe neutrinoless double beta decay, when and if it comes, will need to be followed by corroborating evidence.

On the theoretical side, the early days of neutrino physics featured such luminaries as Wolfgang Pauli, who started it all in 1930, and Enrico Fermi, who gave the first convincing treatment of beta decay in 1934. They were followed by the brilliant Ettore Majorana, who disappeared mysteriously in 1938, and the prophetic Bruno Pontecorvo, who disappeared mysteriously in 1950, only to resurface five years later behind the Iron Curtain.

Neutrino physics continues to be an area in which there is close interaction between theory and experiment. There is, however, a timescale problem. Experiments can take many years between the initial plans and the final results. But theorists operate on timescales of weeks or months. From the fall of 2011, when OPERA announced its (incorrect) result that neutrinos travel faster than light, to the spring of the following year, when they found their mistake, hundreds of theoretical papers poured forth with possible explanations of what was going on.

At present, there is a steady stream of new theoretical ideas on neutrino physics while, at best, the experimental data trickle in. It is almost certain that major new experimental results will be forthcoming, but no one can say which experiment will get there first, or when. Will it be long-distance oscillation experiments like DUNE? Neutrino observatories like IceCube, SNO, or Super-Kamiokande? High-precision measurements like KATRIN, PTOLEMY, or the various neutrinoless double beta decay experiments? New technical advances like coherent neutrino-nucleus scattering? Or perhaps it will be something else entirely. "The most consistent thing about neutrino physics is that it so often surprises us," says Parno. "I don't think the neutrino is out of surprises."

Possibilities abound. Expectations are high. We are waiting for the next act to begin.

Acknowledgments

It's been a tough few years for just about everyone on the planet. We extend a special thanks to all the people who made this book possible in these trying times. We are especially grateful to those who contributed expertise, information, and delightful discussions to our efforts, including Olga Botner, Curtis Callan, Ephraim Fischbach, Joseph Formaggio, Sheldon Glashow, Francis Halzen, Alec Habig, Patrick Huber, Benjamin Jones, Stella Kafka, Mansi Kasliwal, John Learned, William Louis, David Nygren, Erin O'Sullivan, Diana Parno, Rebekah Pestes, Robert Reines, Kate Scholberg, Daniel Stancil, Virginia Trimble, Véronique Van Elewyck, Richard Van de Water, and Anthony Zee.

We thank the four anonymous reviewers who provided insightful comments and suggestions that greatly improved the book. David Kaiser of MIT generously shared his wisdom about the finer details of book authorship. This book would never have happened without the encouragement and professional guidance of the team at the MIT Press, especially Haley Biermann, Deborah Cantor-Adams, and Jermey Matthews. We also appreciate the contributions of Sheila Hill and Stephanie Sakson in preparing the manuscript, and we thank Neta Bahcall, Stephen Parke, and András Patkós who helped in various ways regarding the Lake Balaton conference photo.

Alan Chodos wants to thank his colleagues in the neutrino group at the University of Texas at Arlington, whose active and enthusiastic involvement in DUNE, NeXT, and IceCube was a constant source of inspiration as this book evolved. Equally inspiring was the interest and support displayed by his wife, Florence Haseltine, and coffee companion Doug Kirby, whose contributions made a big difference to him in getting the book over the finish line.

James Riordon is eternally indebted to his partner/colleague/best friend/wife, Martha J. Riordon Heil, for hundreds of edits, thousands of suggestions, and countless hours of discussions about neutrinos and writing. He is also grateful to the late Joseph Weber for sitting for interviews a quarter century ago, and to Project Poltergeist veteran Herald Kruse and his daughter Karen for sharing their neutrino memorabilia. Although his chats with Clyde Cowan were informal, James wouldn't have taken on this book were it not for memories of his grandfather's wit, imagination, and charm. James thanks his sisters, Karen Bradley, Barbara J. Meadows, and Jennifer Hall, along with Uncle George Cowan, for their contributions of family lore and neutrino experiment artifacts. The kids—Nicole, Mary May, J. Patrick, and Marguerite—were sources of inspiration and comfort as we all toiled under the shadow of the global pandemic. A special thanks to our pod family in Los Alamos, Erin Lay, Amy Guthormsen, and their children Charlotte and Walter; and to our dear friends Vicky Mayoral, José Figueroa, and their daughter Olivia, who were crucial companion families during the times of pandemic isolation. James will miss Granny Mary Heil's wisdom and grace and Grandfather Sidney Reed's unbounded kindness, and mourns the many people we lost to COVID and other tragedies during the writing of this book.

Appendix

A.1 The Discovery That Nature Distinguishes Left from Right

As new detection techniques became available, and as physicists returned to their laboratories after the disruptions of World War II, the late 1940s and early 1950s was a time in which many new particles made their appearance. Gone were the halcyon days when it seemed that nature could be understood on the basis of just a few elementary constituents. A whole new menagerie of particles was clamoring for explanation.

Among the most puzzling of these discoveries was a pair of particles that physicists dubbed tau (τ) and theta (θ) (this tau is unrelated to the tau lepton, to be discovered 20 years later, that is now part of the Standard Model). Tau and theta were charged particles, each measured to have the same mass of about 495 MeV/c^2 and the same lifetime. The difference between them was that the τ decayed into three π mesons (or pions), and the θ decayed into two pions.

Pions, or π mesons, were discovered in the late 1940s. They are understood in the Standard Model as bound states of one quark and one antiquark, but of course that was not known in the 1950s when the nature of the tau and theta mesons was being investigated.

The simplest assumption would have been that tau and theta represented two different decay modes of the same particle. But that idea ran into what seemed like an insurmountable roadblock—the state of three pions had different behavior under reflection from the state of two pions, which was impossible if nature respected the symmetry between left and right.

As recognized by Hungarian physicist Eugene Wigner as early as 1927, reflection symmetry in quantum mechanics allows us to endow every state of definite energy with a new label, called parity. Because two reflections are the same as no reflections, this label can take on only two values, +1 or −1. Thus all states of definite energy divide into two categories: those with positive parity and those with negative parity. By the mid-1950s, the pion was known to be a negative parity state.

Reflection symmetry implies that parity is conserved. That is, just as in the case of energy or electric charge, the parity of the state before an interaction takes place must be the same as the parity of the state after the interaction.

A state of two pions has positive parity if its angular momentum is an even multiple of Planck's constant, and negative parity if an odd multiple. A state of three pions is a little more complicated, but under the experimental conditions in which the τ decay was measured, it was possible to deduce that the three-pion state resulting from the decay had negative parity if its total angular momentum was even, and positive parity otherwise. It was, of course, assumed that angular momentum was conserved, so that the pions carried a definite angular momentum, namely, the spin of the parent particle. Therefore, if the θ and τ had the same spin, they had opposite parity, and could not be the same particle, despite all their other similarities.

A session at a physics conference held in Rochester, New York, on the morning of April 7, 1956, brought together many of the physicists who were grappling with the τ-θ puzzle. The session was chaired by Robert Oppenheimer, and the discussion was led by Chen-Ning Yang of the Institute for Advanced Study in Princeton. Yang, together with Tsung-Dao Lee of Columbia University, proposed that there might be a new symmetry of nature that demanded the existence of two particles, of opposite parity, with the same mass. Lee and Jay Orear of Columbia suggested that maybe one of τ and θ was slightly heavier than the other, and could decay into the lighter one by emitting a photon. Robert Marshak of the University of Rochester wondered whether endowing the τ with a higher spin might resolve the paradox.

Finally, Richard Feynman relayed a question from physicist Martin Block, asking if maybe the τ and θ were possibly the same particle, but parity was not conserved. Yang responded, somewhat enigmatically, that he and Lee were looking into it but did not yet have anything to report.

Indeed, Lee and Yang were beginning to think the unthinkable: that parity is not conserved. They carefully examined the experimental evidence and concluded that, although processes involving the strong and electromagnetic interactions definitely conserved parity, there was no such evidence for the weak interactions. They furthermore suggested a number of experiments that could reveal parity violation. Among them, most fatefully, was to examine the beta decay of the cobalt 60 nucleus.

Lee and Yang submitted their paper, coyly titled "Question of Parity Conservation in Weak Interactions," for publication on June 22, 1956. Meanwhile, Lee was also talking to his Columbia colleague, the experimental physicist Chien-Shiung Wu. She immediately appreciated the potential for a revolutionary discovery, and put off a planned trip to Europe and China to perform the cobalt 60 experiment. But she could not do it alone.

The cobalt nucleus, consisting of 27 protons and 33 neutrons, has spin 5. It undergoes beta decay to an excited state of nickel, which then further decays to the nickel ground state via the emission of a pair of photons. The parity-violating quantity to be measured was a correlation between the spin of the nucleus and the momentum of the emitted electron.

Spin (or indeed any angular momentum) and momentum behave oppositely under reflection. To see this, imagine an object that is moving toward a mirror while undergoing rotation given by the right-hand rule (i.e., the rotation is in the direction the fingers of your right hand curl when your thumb points in the direction of motion). The reflected image is moving toward the mirror from the opposite direction, but still rotating in the same direction, so now its rotation is given by the left-hand rule.

Thus any detected correlation would be reversed by reflection in a mirror, providing evidence that parity is violated. But to observe the correlation, Wu had to have a sample of cobalt 60 nuclei that was polarized; that is, the spins had to be preferentially oriented in one direction. If the spin directions were random, the direction of emission of the electrons would be random too, and any correlation would be washed out.

The nuclei could be polarized by placing them in a magnetic field, but the only way to keep them polarized long enough was to do the experiment at extremely low temperatures, well below even that of liquid helium. Wu did not have the necessary expertise, but she was put in contact with a team of physicists from the National Bureau of Standards, led by Ernest Ambler, that did. So Wu traveled to the NBS campus, in Washington, DC, to collaborate with Ambler and his team.

The paper of Wu, Ambler, Hayward, Hoppes, and Hudson was submitted to the *Physical Review* on January 15, 1957. The results were conclusive: Significantly more electrons were emitted antiparallel to the nuclear spin than parallel. The beta decay of cobalt 60 was a parity-violating reaction.

Immediately following the paper of Wu et al. was a paper by Richard Garwin, Leon Lederman, and Marcel Weinrich, also of

Columbia, that also reported the observation of parity violation. It too had been submitted on January 15, but in the first paragraph the authors cede any claim to priority. They had known of Wu's results before doing their experiment.

Garwin et al. used a different reaction, also proposed by Lee and Yang. If parity is violated, then when a pion decays into a muon and a neutrino, the muon can be polarized in its direction of motion (in other words, it is produced with a definite helicity). Garwin et al. measured the polarization by observing the distribution of electrons produced in the subsequent decay of the muon.

To add even more evidence, in the next issue of *Physical Review*, a paper by Jerome I. Friedman and V. L. Telegdi of the University of Chicago also reported the parity violation in pion decay, using a different technique from Garwin et al. Their paper had been received on January 17.

Wolfgang Pauli, who in 1930 had dared to hypothesize the neutrino, was a skeptic about parity violation in 1957. On January 17, before he had heard about the experimental results, he wrote in a letter to Victor Weisskopf of MIT, "I do not believe that the Lord is a weak left-hander, and I am ready to bet a very large sum that the experiments will give symmetric results." But once he learned of the experiments, he was forced to backtrack, writing in a subsequent letter to Weisskopf, "Now, after the first shock is over, I begin to collect myself. Yes, it was very dramatic. On Monday, the twenty-first, at 8 p.m. I was to give a lecture on the neutrino theory. At 5 p.m. I received three experimental papers (those of Wu, Lederman, and Telegdi)." Pauli remarked that he was glad he had not taken the bet, because he could not afford to lose the money. As it was, he said, he had only lost part of his reputation, which he reckoned that he could afford.

We would be remiss if we did not also mention an elegant experiment conducted by Maurice Goldhaber, Lee Grodzins, and Andrew Sunyar of Brookhaven National Laboratory, as described in a paper that was submitted to the *Physical Review* on December 11, 1957. They made use of a very fortuitous reaction in an isotope of the element europium, in which the nucleus absorbs one of the atomic electrons, emitting a neutrino and transitioning to an excited state of the element samarium. The samarium then decays to its ground state by emission of a photon. Preferentially, the photon emerges in the opposite direction as the neutrino, and the crucial point is that, because of angular momentum conservation, given the known spins of the parent and daughter nuclei, the helicity of the photon must be the same as the helicity of the unobserved neutrino. Goldhaber, Grodzins, and Sunyar measured the circular polarization of the photon, thereby determining its helicity and therefore that of the neutrino. Their results were consistent with neutrinos being left-handed 100 percent of the time, evidence for maximal violation of parity in the weak interactions.

The time elapsed between Lee and Yang's suggestion of parity violation and its experimental confirmation by Wu et al. was about seven months. Contrast that with the 25-year interval between Pauli's suggestion of the neutrino and its discovery by Cowan and Reines. In addition, Lee and Yang received the Nobel Prize for their work in 1957, probably the quickest recognition of any achievement in physics by the Nobel committee. As revolutionary as it was, parity violation was immediately accepted by the physics community, and was an important step on the road to the Standard Model.

A.2   The Higgs Mechanism

The final ingredient of the Standard Model, the so-called Higgs mechanism, is a deus ex machina without which the model would fail to describe experimental reality. It is named for Peter Higgs, a British physicist who wrote one of the papers dealing with the mechanism in 1964. The Higgs mechanism is at once incredibly ingenious and embarrassingly contrived.

The main problem that the Higgs mechanism is designed to solve is why the W and Z bosons are so massive, being 86 and 97 times heavier than the proton, respectively. The natural state of a gauge boson is to be massless. The photon is massless, and the gluons are massless too, although, like quarks, they are confined inside the nucleon and its relatives, so we never detect them directly. But the W and Z were detected in the early 1980s, with masses just where the Standard Model predicted them to be on the basis of the known properties of the weak interactions.

The phenomenon of which the Higgs mechanism is an example is called spontaneous symmetry breaking. In this case, the symmetry that breaks is precisely the gauge symmetry for which the W and Z serve as the gauge bosons. The way this happens can be explained by means of an analogy, which like all analogies is not perfect, but hopefully it will illuminate the essence of the problem if the reader is willing to accept it. We begin with a Mexican hat.

Let's place a sombrero on a flat horizontal surface. Not just any sombrero, but a brand-new one, that is fresh from the local haberdashery and still perfectly symmetrical. The symmetry of the hat, namely, rotation about a vertical axis through the crown, is the analogy of the gauge symmetry that is about to break.

We could break the symmetry by giving the sombrero a few well-chosen dents or creases. That is an example of what physicists would call explicit symmetry breaking. Spontaneous symmetry breaking is somewhat more subtle.

We introduce a marble that we balance carefully at the very top of the sombrero. In this configuration, the rotational symmetry is still preserved, and the associated gauge bosons are still massless. But the situation is unstable. The slightest tap, and the marble will roll down the side of the hat, and end up somewhere in the trough around the brim. In that configuration, the marble is stable, but the

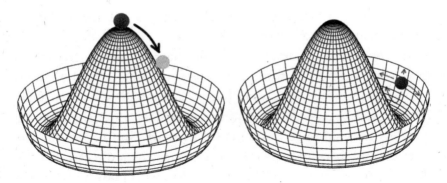

These two figures illustrate some of the aspects of the Higgs mechanism. Before symmetry breaking takes place (left), a marble sits atop the Mexican hat, and the system possesses rotational symmetry about a vertical axis through the crown. But this is unstable; the slightest tap will cause the marble will roll down to the stable configuration in the brim, spontaneously breaking rotational symmetry in the process. Once in the brim (right) the marble can move unimpeded around the trough—this corresponds to massless excitations in the Standard Model. The excitations, when combined with the previously massless W and Z bosons, produce the massive W and Z that are observed experimentally. In addition, as shown, the marble can oscillate up and down the sides of the brim. This corresponds to a massive excitation that survives in the Standard Model, the Higgs boson that was discovered in 2012. *Source:* James Riordon.

rotational symmetry around the vertical axis has been lost, because a rotation moves the marble to a different location along the brim.

With the marble motionless in the trough, it has reached its lowest energy configuration. This is what physicists would call the vacuum, or the ground state. Particles are described by excitations that take the marble out of its ground state. In this case, there are two choices. The marble can move unimpeded around the trough, which corresponds to a massless particle. Or it can try to climb the walls of the brim, which will cause it to oscillate back and forth. This oscillation corresponds to a massive particle. In the real world that the Standard Model describes, there are actually three massless particles, one of which has positive electric charge, one negative electric charge, and the third is neutral. There is a single massive particle, and it is neutral.

What happens next is dictated by the way this contraption interacts with the rest of the Standard Model. The mathematics of its interaction with the W and Z is very dramatic. The two Ws and the Z gobble up the massless particles, and in so doing they transition from their own massless state to a massive one—just what we wanted to happen.

But wait, there's more! The Higgs not only fattens up the W and Z; it also interacts with all the quarks and some of the leptons. The symmetry, before it is broken by the fall of the marble, is so powerful that it forbids any of the quarks and leptons from acquiring mass. Once the symmetry breaks, however, each of these particles acquires a mass that is proportional to the strength of the interaction between it and the Higgs. The notable exceptions are the neutrinos; as long as they are only left-handed, they cannot interact with the Higgs at all, and so even after symmetry breaking they remain massless.

Finally we return to the massive neutral particle that we represented by the marble oscillating back and forth along the sides of the trough. This is a genuine prediction of the Higgs mechanism. It is known as the Higgs boson, and the whole scenario was validated when its experimental discovery was announced at CERN in 2012. Olé!

A.3   An Inside Look at Supernovas

Supernovas are stars that become unstable and explode catastrophically. The physics of supernovas is quite complicated, involving, at various stages, all the interactions of the Standard Model plus gravitation. Not all of this physics is well understood, partly because supernovas are relatively rare events. Although they are expected to occur in our galaxy, the Milky Way, at the rate of a few per century, the last one seen with the naked eye was in 1604, following

Supernova SN1987A produced neutrinos that were the first with origins from outside our solar system to turn up in detectors. *Source:* NASA.

an even brighter one in 1572, but since then there has only been indirect evidence of supernovas in the Milky Way, perhaps because they have been obscured by interstellar dust. Supernova SN1987A occurred not in the Milky Way itself, but in the satellite Large Magellanic Cloud, about 168,000 light years from Earth. So although it was observed here in 1987, the actual explosion had occurred more than 160 millennia earlier.

To maintain a star in equilibrium requires a balance of forces. Gravity always acts to compress, to crush, to make the star collapse under its own weight. Countering this is outward pressure, which comes from two main sources: fusion and what is called *degeneracy pressure.*

During most of its lifetime, the star is powered by nuclear fusion, generating energy as, starting with hydrogen, lighter nuclei fuse to form heavier ones. Each fusion reaction has its own ignition temperature because the nuclei must be moving fast enough to overcome the electric repulsion between them. Only then can they get close enough for the nuclear force to bind them together. The heavier the nuclei, the higher the temperature must be; heavier nuclei tend to have more electric charge than lighter ones.

The heat generated by fusion gives rise to pressure (just as the hot air in a balloon provides enough pressure to keep it inflated). But the nuclear fuel available to the star is limited. The heaviest nucleus that can be produced by fusion is iron, because the iron nucleus is the most stable. Making anything heavier costs more energy than it produces.

Stars of different masses reach different end points, because the lighter stars never get hot enough to go all the way to iron. But eventually (after billions of years), whatever the mass of the star, fusion slows down and comes to a halt. At that point, the main countervailing force against gravity is provided by degeneracy pressure, a consequence of one of the celebrated properties of quantum mechanics, the Pauli Exclusion Principle.

All particles fall into one of two categories, bosons and fermions. In the Standard Model, the gauge bosons and the Higgs boson are (no surprise) bosons, which means that they have spin that is an integral multiple of Planck's constant (recall that gauge bosons have spin 1 and the Higgs is spin 0). Fermions, on the other hand, have spin that is a half-integer multiple of Planck's constant. All the quarks and leptons, including neutrinos, are spin-1/2 fermions.

In addition to the difference in spin, fermions and bosons have very different behavior when it comes to how they fill up the allowed energy levels in a given system. Bosons are gregarious: They like to congregate in the same energy levels. If you try to make an atom out of bosons, they will all happily occupy the lowest energy state.

Fermions, on the other hand, are antisocial. They obey the Exclusion Principle, which forbids more than one of them to occupy the same quantum state. So, in multielectron atoms, the electrons arrange themselves in higher and higher energy levels; once the lower levels are filled, they have no choice. This circumstance gives rise to the rich variety of chemical properties displayed by the elements.

Now imagine a collection of protons and electrons confined in the interior of a star by the star's immense gravity. (Protons, and neutrons, are made up of three quarks and hence they too are fermions.) They cannot be compressed too much, because that would force electrons into higher and higher energy states. Equilibrium is reached when the energy that must be paid to squeeze the electrons further just compensates the energy gained by gravitational compression. The result is an effective pressure that balances gravity.

For stars less massive than the so-called Chandrasekhar limit—about 1.4 times the mass of the sun—when fusion stops, degeneracy pressure holds the core of the star in equilibrium. The resulting object is called a white dwarf. It is very dense: Its mass is about that of the sun, but its size is about that of the Earth. Most white dwarfs do not explode. They just sit there slowly cooling as they radiate their residual heat from fusion into the interstellar medium.

But some white dwarfs continue to accumulate mass after fusion stops. This can happen, for example, if it is bound to a companion star, with mass flowing from the companion onto the white dwarf. As the white dwarf approaches the Chandrasekhar limit, it becomes unstable. It can explode, creating what is classified as a type I supernova. We shall move on, however, to type II supernovas, which occur in more massive stars. This is the class to which supernova SN1987A belongs.

If a star's core exceeds the Chandrasekhar limit, it cannot be stabilized by electron degeneracy pressure. When fusion stops, gravity takes over, causing a runaway collapse and releasing huge amounts of energy in the process. This is the same kind of energy that is used to generate electricity when water tumbles over a waterfall, but of course on a vastly different scale. The energy unit used to describe supernovas is the *foe*, short for "fifty-one ergs," that is, 10^{51} ergs. Supernovas can liberate 100 foe. For comparison, one foe is 27 orders of magnitude larger than the energy of a large hydrogen bomb, so that a supernova is the equivalent of 100,000 trillion trillion H-bombs going off at once.

Most of this energy, about 99 percent, is released in the form of neutrinos. The main reason for this is simply that the stellar core is so dense that nothing can escape, except for the weakly interacting neutrinos. In the initial phases of the collapse it becomes energetically favorable for the electrons to combine with protons to create neutrons and release neutrinos, kind of an inverse of beta decay, which acts to reduce the degeneracy pressure.

Normally, of course, this can't happen because the neutron is heavier than a proton and an electron combined. That's why a hydrogen atom won't disappear due to the proton in the nucleus gobbling up the orbiting electron. But in the dense stellar core, energy can be gained when a proton captures an electron from a high-energy state and turns into a neutron. As the collapse proceeds and the temperature rises, neutrinos from this process are augmented by thermal radiation of neutrino-antineutrino pairs, mediated by Z-bosons that are emitted by the highly excited particles in the core. Most of the energy, in fact, is emitted by these thermally radiated neutrino-antineutrino pairs.

The ultimate fate of a type II supernova depends on its mass. If it is not too heavy, a denser configuration can be supported by degeneracy pressure of neutrons alone. As the collapse proceeds, the protons keep swallowing electrons and turning into neutrons. But the neutrons themselves are fermions, so they resist the gravitational compression. Furthermore, the strong interaction provides a repulsive force between neutrons at short distances. So the end result may be a stable configuration, dubbed a neutron star, that can be thought of as roughly a gigantic atomic nucleus, made up entirely of neutrons. Current calculations indicate that such a configuration can be stable against gravitational collapse provided that the mass of the stellar core that remains after the explosion is less than about twice the mass of our sun. The characteristic size of such an object would be about 10 kilometers. It is thought that the core of super-nova SN1987A ended as a neutron star.

If the mass of the core is more than twice the mass of our sun, various even more exotic and dense configurations might be possible, such as a quark star, which would resemble not a gigantic nucleus but a gigantic nucleon, composed of quarks of various flavors. Whether or not such things exist, current theory asserts, and observation increasingly confirms, that there is a threshold mass above which a stellar core will collapse beyond rescue into a gravitational black hole.

Glossary

Antimatter: Corresponding to every particle there is an antiparticle. This is both an experimental observation and a theoretical consequence of the combined principles of quantum mechanics (q.v.) and special relativity (q.v.). The first antiparticle to be discovered was the positron, antiparticle to the electron (q.v.). Antiparticles and particles have exactly the same mass, and, if they are unstable, the same lifetime, but they have opposite values of other attributes like electric charge. A typical example is the triplet of pi-mesons (or pions), which are bound states of a quark and an antiquark. The positively and negatively charged pions are a particle-antiparticle pair, with exactly the same mass. The neutral pion is its own antiparticle, and is slightly lighter than the other two. The photon is its own antiparticle, as is the Z boson, while the W+ and W− bosons are a particle-antiparticle pair. The neutrinos are the only uncharged spin-1/2 particles in the Standard Model, and the question of whether they are their own antiparticles or not is a subject of active investigation. At the level of elementary particles, matter and antimatter appear on very much the same footing, but physicists are trying to exploit the small lack of symmetry between them to explain why the universe as a whole is made of one and not the other.

Atomic bomb: There are two kinds, the regular and the extra-powerful. The regular kind is based on nuclear fission (q.v.). That is, energy is released when a nucleus, such as uranium or plutonium, splits apart into the nuclei of two lighter atoms after being struck by a neutron. Not only is energy released, but more neutrons are as well, which strike more nuclei, generating a chain reaction and creating a powerful explosion. Bombs based on uranium and plutonium fission were first made under the aegis of the Manhattan Project in World War II, and were dropped on the Japanese cities of Hiroshima (uranium) and Nagasaki (plutonium). The extra-powerful bombs add nuclear fusion (q.v.) to the explosion, in which hydrogen nuclei fuse to form helium, releasing even greater amounts of energy than is available from fission. Nuclear fusion is the energy source that powers the stars. Fusion bombs were developed in the 1950s in both the United States and the Soviet Union.

Axion: A hypothetical particle that may be a significant component of dark matter (q.v.). Axions made their theoretical debut in the context of a possible explanation of certain symmetries of the strong interactions. Unlike most other particles with funny appellations, their name, inspired by a laundry detergent, was given to them not by Nobel Prize–winning physicist Murray Gell-Mann (q.v.), but by Nobel Prize–winning physicist Frank Wilczek. Particles called axion-like particles (ALPs), with many of the same properties, crop up in other contexts not related to the strong-interaction symmetry problem. Axions and ALPs have been widely sought, in everything from liquid xenon to microwave ovens, but so far without success.

Boson: In the quantum world, all particles have intrinsic angular momentum called spin (q.v.), and spin is quantized to be either an integer or a half-integer multiple of a fundamental constant of nature called Planck's constant (q.v.). Particles whose spin is an integer multiple of Planck's constant are called bosons, named after the Indian physicist Satyendra Nath Bose (unrelated to the maker of audio equipment, so far as is known). The half-integer variety are called fermions, after the Italian physicist Enrico Fermi. Familiar examples are photons (spin-one bosons) and electrons (spin-1/2 fermions). Bosons and fermions differ not only in spin, but also in their social behavior. Bosons are gregarious—they like to congregate in the same quantum state. That's why a laser works so well. Fermions are stand-offish. Once a fermion occupies a state, no other fermion is allowed in. That's why electrons build up complex atoms and molecules.

Cherenkov radiation: Nothing travels faster than light—in a vacuum. But in a material medium, light slows down; for example, in water it moves at only three fourths its vacuum speed. Under these circumstances, it's possible for individual particles to move faster than light inside the medium, as long as the particles don't exceed the vacuum speed of light. If the particle is charged, it will emit electromagnetic radiation, called Cherenkov radiation after the Soviet physicist Pavel Cherenkov, who first observed it in the 1930s. The phenomenon is the analog for light of the sonic boom produced by an airplane traveling through air faster than the speed of sound, or to the "V"-shaped wake of a boat traveling faster than the speed of the water waves. Cherenkov radiation plays a crucial role in water-based neutrino detectors such as Super-Kamiokande, which contains ultrapure water in a tank surrounded by light-detecting photomultiplier tubes. When a sufficiently energetic neutrino scatters in the water, it produces a charged lepton moving faster than light, and the Cherenkov radiation is detected. The angle of the radiation (think of the angle made by the "V" of a boat's wake) reveals the speed of the particle, enabling researchers to tell which flavor it is, and also the direction from which it came. Another example of Cherenkov radiation is the eerie bluish glow in water-cooled nuclear reactors, produced by fast-moving charged particles that are created in the fission process.

Chirality: Chirality means "handedness." Objects that possess chirality come in two varieties, right- and left-handed. There are many macroscopic examples, most obviously, your hands, and also screws, which can be right-handed (common) and left-handed (not so common). A right-handed screw will advance in the direction of the thumb of your right hand when twisted in the direction that the fingers of your right hand are curling. And correspondingly for the left-handed screw. There are lots of molecules that come in right- and left-handed varieties too. In all cases, if you view a right-handed object in a mirror, it looks like its left-handed version, and vice-versa. Important for us is that spin-1/2 elementary particles also possess chirality. They come in right- and left-handed versions. But if the particle has mass, this chirality is not preserved in time. A spin-1/2 particle that starts out right-handed will evolve, as time goes on, into a superposition of right- and left-handed.

Classical mechanics: In the late seventeenth century, Isaac Newton formulated his laws of motion, which form the basis of classical mechanics. Perhaps most familiar is the statement that force equals mass times acceleration (or $F = ma$), probably the best known equation in physics at least until $E = mc^2$ came along. Classical mechanics is deterministic: If you are given the positions and velocities of a collection of particles at a particular time (say, $t = 0$), and you know the forces acting on those particles, then you can determine their positions and velocities any time in the future (or in the past, for that matter). It came as a great shock when physicists realized, in the early twentieth century, that classical mechanics no longer worked in the atomic and subatomic regime.

Dark energy: Something very strange is happening in the farther reaches of the universe. You might think, since gravity is an attractive force, that the galaxies and clusters of galaxies that have been hurtling away from each other since the Big Bang would be slowing down as gravity tries to reel them back in. But no. Measurements reported in the late 1990s found that the expansion of the universe is accelerating, not slowing down. Something is pushing everything apart. Physicist Michael Turner gave it a name, dark energy, but what it is nobody, not even Turner, knows. What we do know is that dark energy constitutes about 68 percent of the energy contained in the observable universe. Another 27 percent is so-called dark matter (q.v.). Ordinary matter, which makes up you and me and everything we are familiar with, is only the remaining 5 percent.

Dark matter: There is evidence, both at the level of galaxies and at the level of clusters of galaxies, that more mass is needed, beyond the ordinary matter that is detectable, to hold them together. We know this so-called dark matter must contain about five times more energy than ordinary matter. It behaves like ordinary matter gravitationally, and interacts in other ways with ordinary matter very weakly, if at all. In particular, it must be dark, meaning it does not absorb or emit photons. It also must be cold, meaning that when the galaxies and galactic clusters were being formed, the dark matter was moving slowly. This is necessary to reproduce the pattern of structure formation that is observed. If the dark matter were too hot, the universe we see today would not be clumpy enough. What makes up dark matter is not currently known; some possibilities are axions (q.v.) and WIMPs (q.v.), but there are many other potential candidates, and the ongoing experimental search for dark matter has so far come up empty.

Dirac particle: A Dirac particle is a spin-1/2 fermion that comes with a distinct anti-particle. Both particle and antiparticle have right-handed and left-handed varieties. A Majorana particle, in contrast, is a single left-handed/right-handed pair, and is its own antiparticle. A particle with nonzero electric charge is always a Dirac particle, because the antiparticle must have opposite electric charge and cannot be related to the particle simply by changing its handedness.

Electromagnetic spectrum: In the 1860s, Scottish physicist James Clerk Maxwell unified all electric and magnetic phenomena with a set of equations that predicted electromagnetic waves, and that moreover predicted that these waves traveled at a speed that equaled the measured speed of light, so, as a bonus, his theory showed that light itself was an electromagnetic wave. Electromagnetic waves can have any wavelength, from zero to infinity. The visible part of the spectrum, light, is a small piece, ranging in wavelength from 380 billionths of a meter (violet) to 700 billionths of a meter (red). It is no coincidence that the spectrum of radiation put out by the sun peaks in the visible range. In 1905 Einstein proposed that electromagnetic radiation also has particle properties, being composed of what later were called photons. To every wavelength there corresponds a photon of a particular energy, proportional to Planck's constant (q.v.) divided by the wavelength. Thus, for example, red photons have lower energy than violet photons, because red light has a longer wavelength than violet light. Different parts of the spectrum have their own names. On either side of the visible are infrared (longer wavelengths) and ultraviolet (shorter). Beyond infrared are radio waves and microwaves; beyond ultraviolet are X-rays and gamma rays, with many other divisions and subdivisions in between.

Electron: The electron was the first of what we now think of as elementary particles (q.v.) to be discovered, by the English physicist J. J. Thomson in 1897. The electron is a key component of atoms, since every neutral atom consists of a nucleus with Z protons (and some number of neutrons), surrounded by a cloud of Z electrons. (Following standard physics notation, we use "Z" to label the number of protons in a nucleus. This should not be confused with the Z boson, one of the gauge bosons of the Standard Model.) Conventionally, the proton has an electric charge of +1, and the electron of –1. All other electric charges are integer multiples of those. The electron is the lightest charged particle, with a mass almost 2,000 times smaller than the proton's. It is typically the charge carrier of electric current, and because its charge is negative, if current flows to the left, the electrons are actually moving to the right, to the confusion of generations of physics students (we have Benjamin Franklin to thank for this). In the Standard Model, the electron is the lightest of the three charged leptons, and is accompanied by its own flavor of neutrino. The electron interacts both electromagnetically and via the weak force. Like all particles, it has an antiparticle, the positron, discovered in 1932, which has exactly the same mass but opposite electric charge. The electron and positron can form a bound state, positronium, in which the particles orbit each other held together by their electric attraction, until they eventually annihilate into two or three photons.

Elementary particle: Intuitively, an elementary particle is a building block that goes into making up other stuff, while not itself being made of anything smaller. What qualifies as "elementary" depends on time and on the discipline. The ancients conceived of air, earth, fire, and water as basic constituents. Atoms of the chemical elements that appear on the periodic table are taken as elementary in much of chemistry. But, of course, atoms are made up of smaller things. With the great success of the Standard Model physicists are tempted to think of the roster of particles appearing therein as elementary. Indeed, the electron has been probed to distances of about 10^{-17} centimeters, and no underlying structure has been found. But the Standard Model leaves out gravity, and one can combine the gravitational constant, which appears in Newton's law of gravitation, together with Planck's constant (q.v.) and the speed of light, to form a fundamental constant with the dimensions of length, called the Planck length. Its value is about 10^{-33} centimeters, so perhaps we have 16 orders of magnitude to go before we can confidently say that particles we now think are elementary really are. What are the chances of that?

eV (electron volt): The voltage of a battery is a measure of its ability to produce current, which really means to accelerate electrons. If you place an electron (q.v.) at the cathode, or negative terminal, of a one-volt battery, and let it flow to the anode, or positive terminal, when it gets there the electron will have increased its energy by one electron volt. So the electron volt is a measure of energy, one that is convenient for atomic physics because the binding energies of electrons in atoms are typically tens to hundreds of electron volts. Other convenient measures of energy are the MeV (million electron volts) because the rest-mass energy of the electron is about half an MeV, and the giga electron volt (GeV, a thousand MeV), because the rest-mass energies of the proton and neutron are each just under a giga electron volt. By contrast, neutrino rest-mass energies, when finally measured, are expected to be small fractions of an electron volt.

Exchange force: In classical mechanics (q.v.), forces are usually just input information that is inserted into the equation $F = ma$ to solve for the motion of a particle. In the quantum world, and in particular in the Standard Model of particle physics, what is given is not the force but rather the rule for calculating how a spin-1/2 particle, such as the electron, either emits or absorbs a spin-one boson, such as the photon.

Then the force between two electrons is generated by the first electron emitting a photon and the second electron absorbing it. Such a force is called an exchange force, and the same mechanism applies also to the other particles of the Standard Model. For example, quarks interact strongly with each other by exchanging gluons, and the neutrinos, which feel only the weak force, interact with the other particles in the Standard Model by exchanging the W and Z bosons.

Fermion: See Boson.

Fission: Radioactive decay of nuclei was discovered in the late nineteenth century, and classified into three main types. In each of them, the nucleus emits a small particle, and doesn't move very far in the periodic table (in beta decay, it moves up or down one; in alpha decay, down two, and in gamma decay it stays put). But nuclear fission is quite different. Some very heavy nuclei are so bloated that they are close to being unstable. Their protons repel each other due to their electric charge, while the short-range nuclear force, which is holding everything together, becomes less effective for bigger nuclei. Along comes a stray neutron, gives the nucleus a tap, and splits it into two roughly equal pieces, with the release of energy and a few smaller shards including some extra neutrons. After its experimental discovery by Hahn and Strassmann in Germany in late 1938, and its interpretation by Frisch and Meitner in Sweden as nuclear fission (Lise Meitner had been Hahn's longtime collaborator, but she was forced to flee Germany because of the Nazis), its potential use in bomb-making was quickly realized. See Atomic bomb.

Fundamental particle: See Elementary particle.

Fusion: The curve of binding energy is very interesting. The nucleus of iron, an element in the middle of the periodic table, is the most tightly bound. As one goes in the direction of heavier nuclei, the binding decreases, which means that nuclear fission (q.v.) of heavier nuclei releases energy. On the other side, combining nuclei lighter than iron, that is, nuclear fusion, also releases energy. In fact, the curve is steeper on the lighter side, so fusion is actually more efficient at generating energy than is fission. The problem with fusion, however, is that it is hard to achieve. The extra binding comes from the short-range nuclear force, but before it kicks in, one must overcome the long-range repulsion of the positively charged nuclei. Controlled fusion means keeping very hot atoms in close proximity, so that their nuclei can bang into each other at high speeds, overcome the electric repulsion, and fuse into heavier nuclei. Stars use gravity to confine the nuclei, but here on earth the problem is much more challenging—unless one wants to have uncontrolled fusion, aka a hydrogen bomb. See Atomic bomb.

Gauge boson: See Boson; Exchange force; Gauge symmetry.

Gauge symmetry: A gauge symmetry is an example of a symmetry (q.v.) with the special property that the transformation characterizing the symmetry can be implemented independently at each point of space and time. This is not true of obvious symmetries like translation invariance: If you have an experimental apparatus consisting of a source and a detector, for example, you have to move them both together so as not to affect the outcome; you can't move the source down the hall and the detector into the basement and expect to get the same result. But certain internal symmetries do have this property. The price one has to pay for gauge symmetry, however, is that the way particles interact with each other is very tightly constrained. In the Standard Model, its gauge symmetry dictates that all the spin-1/2 particles must interact with spin-one gauge bosons, which carry the force between the spin-1/2 particles (see Exchange force). The fact that symmetry determines (almost) all the interactions in the Standard Model is one of its most attractive features.

General relativity: Ten years passed between the appearance of special relativity (q.v.) in 1905 and general relativity in 1915. During much of that time, Einstein was struggling to incorporate gravity into his theory. He was guided by the principle of equivalence, which says that in a sufficiently small region, the effects of gravity can be eliminated by acceleration (that's why orbiting astronauts, who are in free-fall around the Earth, are weightless). General relativity is a geometrical theory. The presence of matter causes space-time to curve, and the curvature of space-time dictates how the matter will move. Most physicists would agree that it is the most beautiful theory ever constructed. It is also among the most successful, having passed numerous tests where it predicts deviations from Newton's law of universal gravitation. These successes continue today, with the observation of gravitational waves from colliding black holes, strictly in accord with general relativity. But Einstein famously did not like quantum mechanics, and his greatest creation doesn't, either. The problem of uniting general relativity with quantum mechanics remains a challenging problem for theoretical physics.

Hadron: Hadrons are particles that feel the strong force. According to the Standard Model, they are made up of quarks. One subset, called baryons, contains three quarks, the simplest examples being the proton (q.v.) and the neutron (q.v.). Another subset, called mesons, are made up of quark-antiquark pairs, the simplest example being the pi-meson or pion, which comes in three varieties with electric charges +1, −1, and 0. As the lightest hadron, the pion is copiously produced in particle accelerators.

Handedness: See Chirality.

Helicity: For massive particles, spin (q.v.) can be measured by bringing the particle to rest and seeing how much angular momentum it has. But massless particles are never at rest, so the spin is defined differently: It is the angular momentum in the direction of its motion, which is called the helicity. The helicity can be positive, if the angular momentum points along the direction of motion, or negative, if it points in the opposite direction. For a massless spin-1/2 particle, the helicity and chirality (q.v.) are the same, with positive helicity being equal to right-handedness, and vice versa. For a massive spin-1/2 particle, the connection is no longer exact, with positive and negative helicity each being a mixture of right- and left-handed.

KeV (kilo electron volt): A thousand eV (q.v.) or a thousandth of an MeV.

Lambda CDM: The currently favored standard model of cosmology goes by this moniker. The lambda stands for dark energy (q.v.) and the CDM is short for cold dark matter (see Dark matter).

Lepton: The leptons are the spin-1/2 Standard Model particles that do not feel the strong force. They come in three flavors, each flavor comprising one lepton of charge −1, and an accompanying neutrino (q.v.) with zero electric charge. The charged leptons, in increasing order of mass, are the electron, the muon, and the tau, and the neutrinos are named according to their charged partners.

Lorentz invariance: Hendrik Antoon Lorentz was an eminent Dutch physicist in the late nineteenth and early twentieth century. He worked on some of the same problems that motivated Einstein's special theory of relativity in 1905. Special relativity (q.v.) informs us that observers moving relative to each other with constant velocity experience space and time intervals differently. The rules that express the space and time intervals according to one observer in terms of the space and time intervals of a second observer moving relative to the first are known as Lorentz transformations. Invariance under these transformations is the symmetry (q.v.) associated with any system that respects the laws of the special theory of relativity.

Magnetic moment: If a couple on their first date each looks lovingly into the other's eyes, are they having a magnetic moment? Yes, but that's not what we're talking about here. For a particle, the magnetic moment is a measure of its strength of interaction with an external magnetic field. A spinning charge behaves like a little bar magnet, so one expects an electron or muon, each of which has an electric charge and spin 1/2, to possess a magnetic moment. The neutrino has spin, but no charge, so how can it have a magnetic moment? Due to quantum effects, the neutrino can turn itself into a W boson and an electron for a very short time, and then turn back into a neutrino again. In that brief interval it can interact with the external field. This will, however, be a very small effect, but potentially a very interesting one, because its size could reveal beyond-the-standard-model physics, and its detailed nature could distinguish between Dirac particles (q.v.) and Majorana neutrinos. So far, no electromagnetic properties of neutrinos have been detected.

Majorana particle: See Dirac particle.

MeV (million electron volt): See eV.

Murray Gell-Mann: Murray Gell-Mann was a highly influential theoretical particle physicist who bestrode the field from the late 1950s through the 1960s. He is known not only for his discoveries but for the names he bestowed on them. "Strangeness" arrived in the 1950s, to be followed by "quark," "color," "flavor," and perhaps a few more. His cachet inspired others to similar flights. Before Gell-Mann there were "proton" (q.v.), "neutron" (q.v.), and "electron" (q.v.). Afterward, "charm," "beauty," "axion" (q.v.), "WIMP" (q.v.), and "Loryon." One must not let the fanciful names of these entities disguise the seriousness with which physicists pursue their theoretical and experimental implications.

Negative energy: In the absence of gravity (which is neglected in the Standard Model), only energy differences have physical significance—there is no absolute way to fix the zero of energy. For example, it is often said that the ground-state energy of the hydrogen atom is -13.6 eV, a negative number. But that is only because the zero of energy was arbitrarily chosen to be where the electron (q.v.) leaves the proton (q.v.) and becomes unbound. The real significance of -13.6 eV is that the rest mass energy of the hydrogen atom is less than the sum of the rest mass energies of its constituent electron and proton by that amount. Likewise, the rest mass of a nucleus is less than the sum of the rest masses of its constituent protons and neutrons (q.v.) by an amount called the binding energy, which is negative. A problem does occur if the amount of negative energy is not bounded from below, that is, if there is no "floor." Then the system is unstable and will cascade downward into the abyss. This is the problem classical mechanics (q.v.) faced with regard to the hydrogen atom: The orbiting electron would spiral ever closer to the proton and radiate energy indefinitely, whereas quantum mechanics (q.v.) predicts a stable ground state. It's also the problem Dirac faced with his equation for the electron: The negative energy solutions were not bounded below, and Dirac postulated the filled negative energy sea to explain how the system could be stable.

Neutrino: The lightest spin-1/2 particle in the Standard Model. It has no electric charge and comes in three flavors—electron, muon, and tau—which are superpositions of three states of definite mass, giving rise to the phenomenon of neutrino oscillations (q.v.). Neutrinos feel neither the strong nor the electromagnetic force, and interact only via the weak force, by exchanging the W and Z gauge bosons (q.v.).

Neutrino oscillations: Neutrinos come in three flavors, each of which is a particular superposition of three states of definite mass. According to quantum mechanics, a neutrino starting off as one flavor will become a superposition of all three flavors as it propagates, meaning that in a subsequent measurement it could show up as a different flavor from the initial one. There are many subtle aspects to this phenomenon, and many complications, including the fact that solar neutrinos, for example, travel through a dense background of electrons on their way out of the sun, which significantly affects their oscillations.

Neutron: One of two nuclear constituents, the other being the proton. The neutron, as its name suggests, has zero electric charge. The position of an atom on the periodic table is determined by the number of its protons, whereas its mass is determined by the number of protons and neutrons together. Atoms with the same number of protons but different number of neutrons are isotopes of each other, with essentially the same chemical properties. Since protons repel each other due to their electric charge, as nuclei get heavier, the proportion of neutrons tends to increase. Like protons, neutrons are not elementary, being composed of three quarks held together by gluons. In free space, the neutron can decay, converting to a proton, an electron, and an electron antineutrino, but in stable nuclei this process cannot occur, because the energy gained by converting the neutron to the lighter proton is more than offset by the extra energy needed to replace a neutron by a proton inside the nucleus.

Nucleon: "Nucleon" is physicists' shorthand for either a neutron (q.v.) or a proton (q.v.). If you want to talk about a constituent of a nucleus without caring which one it is, you just say "nucleon."

PeV (peta electron volt): A million GeV (see eV). Particles produced by accelerators on earth are a factor of at least a hundred below PeV energies, but cosmic rays are known with PeV energies and higher. Not to be confused with PEV, or a plug-in electric vehicle.

Photon: See Electromagnetic spectrum.

Planck's constant: When a new theory comes along, it often brings with it a new fundamental constant of nature. Newton's law of universal gravitation came with the gravitational constant that determines how strong gravity is. The theory of relativity introduced the speed of light as a constant. Likewise, quantum mechanics requires a new constant, first used by the German physicist Max Planck in 1900 and quite appropriately known as Planck's constant. It is very small on macroscopic scales of size and energy, and hence we do not notice quantum effects directly in everyday phenomena. But once we get down to atomic sizes and smaller, the situation changes. The spins of elementary particles are quantized in units of Planck's constant, and the uncertainty in a particle's position times the uncertainty in its momentum, as embodied in Heisenberg's famous principle, is proportional to Planck's constant. In a world where Planck's constant is set to zero, we can know a particle's position and velocity exactly, just as classical mechanics (q.v.) demands. But in the subatomic world, where Planck's constant is significant, quantum effects cannot be ignored.

Proton: The proton is one of two types of particle that make up the atomic nucleus, the other being the neutron (q.v.). The proton carries the fundamental unit of electric charge, of which any other electric charge is an integer multiple (positive, negative, or zero). The rest-mass energy of the proton is a little under one GeV, slightly smaller than that of the neutron. The proton is not elementary; inside are three quarks (q.v.) held together by gluons. As far as is currently known, the proton is absolutely stable, although according to a number of theories it is actually unstable, but with a very long lifetime. Either way, none of us has to worry that our protons will decay before we do.

Quantum mechanics: When physicists discovered that classical mechanics (q.v.) could not explain atomic or subatomic phenomena, they managed, by 1925–1926, to construct a new set of rules, quantum mechanics, that successfully replaced it. But the price was high: Quantum mechanics is no longer deterministic, dealing instead in probabilities, and it endows the microscopic world with attributes that seem weird and counterintuitive, such as wave-particle duality, superposition, and entanglement, leading Richard Feynman to proclaim, famously, that nobody understands it. There are a number of competing interpretations, such as the mainstream Copenhagen interpretation and the ever-popular many worlds interpretation; there are also those who seek a deeper level of reality that would account for quantum phenomena. But the silent majority of physicists belong to the "shut up and calculate" school, according to which you follow the rules, get the answer, and compare with experiment, without stopping to ask what it all really means.

Quark: Quarks are Standard Model spin-1/2 particles that interact with gluons, the force-carriers of the strong force, which is so strong that quarks, together with the gluons, are permanently bound inside hadrons (q.v.). Quarks come in three colors, color being the analog of electric charge for the strong force. They are unusual in that their charges are either 2/3 or −1/3 times the fundamental charge, but they are always combined in such a way that the hadrons have integer multiples of the fundamental charge. Quarks also come in six flavors, grouped into three generations: (up, down), (charm, strange), and (top, bottom). In addition to every quark there is a corresponding antiquark. Physicists disagree among themselves as to whether "quark" rhymes with "park" or "pork," and intensive research reveals that a good case can be made either way.

Special relativity: Formulated by Einstein in 1905, special relativity deals with the relationship between observers moving with constant velocity relative to each other. The postulates are that (1) there is no preferred observer, that is, no one can claim to be "absolutely" at rest; and (2) the speed of light is the same for all observers. From these one deduces some remarkable consequences, such as that an unstable particle moving through the laboratory will take longer to decay than one at rest (time dilation) and that moving objects appear foreshortened in their direction of motion (length contraction). Special relativity also provides a relation between a particle's momentum and energy that includes not only its kinetic energy, or energy of motion, but also its rest mass energy through Einstein's famous equation $E = mc^2$.

Spin: Angular momentum is a familiar phenomenon. Macroscopically, it always involves rotation about an axis. For example, if you tie a rock to one end of a rope, and swing it in a horizontal plane about your head, then the amount of angular momentum is proportional to three things: the mass of the rock, its distance from the axis (i.e., the length of the rope), and the rate at which it is swung. (The direction of angular momentum is the direction the thumb of the right hand points if the fingers are curled in the direction that the rock is rotating.) Elementary particles, obeying the rules of quantum mechanics, can have angular momentum of this type, called orbital angular momentum, which is carried, for example, by certain electrons bound in atoms. But elementary particles also have a second, intrinsic form of angular momentum, called spin, that persists even if they are not moving. Like all angular momentum in quantum mechanics, it always comes in either integer or half-integer multiples of Planck's constant (q.v.). It is necessary to include the spin because only total angular momentum, involving both orbital and spin components, is conserved. Particles with integer spin are called bosons (q.v.); those with half-inter spin are called fermions (q.v.).

Sterile neutrino: Sterile neutrinos may or may not exist, but if they do, it's because the weak interaction involves only left-handed particles and right-handed antiparticles (see Chirality). Therefore, any right-handed neutrino, or left-handed antineutrino, does not feel the weak interaction. In fact, it doesn't feel any known interaction, except possibly gravity, which is far too weak to show up in particle physics experiments. So the only way we would know that sterile neutrinos exist is if ordinary neutrinos, which can mix with them, perform a disappearing act. Certain short-baseline oscillation experiments (i.e., carried out over tens or hundreds of meters) could be explained by invoking sterile neutrinos of the appropriate mass, but other experiments that should have seen evidence of them instead seem to rule them out.

Symmetry: Theoretical physicists love symmetry, because it adds elegance and beauty, and also allows them to better understand what's going on. A physical system possesses a symmetry if you can transform it in some way without affecting its physical properties. For example, translation symmetry means you can move your apparatus to the laboratory down the hall and still get the same experimental result. Time translation symmetry means you can do the experiment now or next Saturday and still get the same result. Rotation symmetry means your detector can point east or it can point 10 degrees north by northeast and you will get the same result. A bonus is provided by Noether's theorem, named for German mathematician Emmy Noether: To every symmetry is associated a conservation law. If your theory is translation invariant, momentum is conserved. If it is time-translation invariant, energy is conserved. And rotational invariance implies angular momentum conservation. In addition to these easily visualized symmetries, theories often possess internal symmetries, in which particles transform among themselves in some way while leaving the physics invariant. Examples of these are gauge symmetries (q.v.).

TeV (tera electron volt): A thousand GeV (see eV). The Large Hadron Collider (LHC) at the CERN laboratory in Switzerland, currently the most powerful particle accelerator, operates in the 10 TeV range.

WIMP (weakly interacting massive particle): WIMPs held center stage for many years as the leading candidates for dark matter (q.v.), primarily because these particles, predicted by several beyond-the-standard-model scenarios, had just the right masses and interaction strengths that their density today would coincide with what dark matter requires. This coincidence was referred to as the "WIMP miracle." Unfortunately, all experimental searches for WIMPs have been negative, to the point where now the limelight has shifted to other candidates, including axions (q.v.).

Notes

Portrait of a Ghost

1. John Updike, "Cosmic Gall," in *Collected Poems 1953–1993* (New York: Knopf, 1993), 313.

Chapter 1

1. Isaac Asimov, *The Neutrino* (Garden City, NY: Doubleday & Company, 1966), xiii.

2. Pronounced "RY-nəs."

3. Virginia Trimble and Frederick Reines, "The Solar Neutrino Problem—A Progress (?) Report," *Reviews of Modern Physics* 45, no. 1 (1973): 1.

4. Murray Gell-Mann, "Beauty, Truth and . . . Physics?," TED2007, March 2007. www .ted.com/talks/murray_gell_mann_beauty_truth_and_physics?language=en.

Chapter 2

1. H. Richard Crane, "The Energy and Momentum Relations in the Beta-Decay and the Search for the Neutrino," *Reviews of Modern Physics* 20, no. 2 (1948): 278.

2. A few years later, Rutherford proved that alpha particles are helium nuclei by capturing many of them and showing that they had the same properties as helium once they had electrons attached.

3. Lise Meitner, "On the Origin of Beta-Decay Spectra of Radioactive Substance," *Zeitschrift für Physik* 9 (1922): 131–145.

4. R. C. Tolman, "Remarks on the Possible Failure of Energy Conservation," *Proceedings of the National Academy of Sciences of the United States of America* 20, no. 6 (1934): 379–383, doi:10.1073/pnas.20.6.379.

5. Emphasis added.

6. Charles P. Enz, *No Time to Be Brief: A Scientific Biography of Wolfgang Pauli* (Oxford: Oxford University Press, 2010).

7. The authors thank Professor Donna Hoffmeister for assistance with the translation from the German.

8. James Chadwick, "Possible Existence of a Neutron," *Nature* 129, no. 3252 (1932): 312.

9. Enrico Fermi, "An Attempt to a β Rays Theory," *Il Nuovo Cimento* 1 (1934): 20.

10. F. Perrin, "Natural Particles of Intrinsic Mass 0," *Comptes Rendus* 197 (1933): 1625.

11. I. Curie and F. Joliot, "Artificial Production of a New Kind of Radio-Element," *Nature* 133, nos. 201–202 (February 10, 1934).

12. H. Bethe and R. Peierls, "The Neutrino," *Nature* 133 (1934): 689–690.

13. C. Sutton, *Spaceship Neutrino* (New York: Cambridge University Press, 1992), xi.

14. M. Nahmias, "An Attempt to Detect the Neutrino," *Mathematical Proceedings of the Cambridge Philosophical Society* 31, no. 1 (1935): 99–107. doi:10.1017/S0305004100012986.

15. H. R. Crane, "An Attempt to Observe the Absorption of Neutrinos," *Physical Review* 55 (March 1, 1939): 501.

Chapter 3

1. W. Kropp, M. Moe, L. Price, J. Schultz, and H. Sobel, *Neutrinos and Other Matters* (Teaneck, NJ: World Scientific, 1989), 1.

2. From private conversations with James Riordon.

3. Kropp et al., *Neutrinos and Other Matters*, 554.

4. Kropp et al., *Neutrinos and Other Matters*, 554.

5. C. L. Cowan Jr., F. Reines, F. B. Harrison, H. W. Kruse, and A. D. McGuire, "Detection of the Free Neutrino: A Confirmation," *Science* 124, no. 3212 (July 20, 1956): 103–104.

6. B. Pontecorvo, "Inverse Beta Process," National Research Council of Canada, Division of Atomic Energy, Chalk River, 1946, Report PD-205, http://131.114.73.131/Articles/Inverse_beta_process_Report_PD205-1946.pdf.

7. Raymond Davis Jr., "Remembering Clyde Cowan," *AIP Conference Proceedings* 52, no. 1 (1979): 15.

8. Raymond Davis Jr., Nobel Lecture. NobelPrize.org. Nobel Prize Outreach AB 2021, October 13, 2021.

9. Nomination Archive, NobelPrize.org. Nobel Prize Outreach AB 2021, September 27, 2021.

10. Several of Reines's reports on the effects of nuclear explosions were declassified in 1996, the year after he collected his Nobel Prize.

11. Frederick Reines—Nobel Lecture, NobelPrize.org. Nobel Prize Outreach AB 2021, October 12, 2021.

Chapter 4

1. Brian Duignan, "Occam's Razor," in *Encyclopedia Britannica*, May 28, 2021, https://www.britannica.com/topic/Occams-razor, accessed February 24, 2022.

2. Pronounced "mai-your-RA-na."

3. Negative energy in and of itself was not a problem—many established physics theories include negative energy states.

4. Richard Feynman, *QED: The Strange Theory of Light and Matter* (Princeton, NJ: Princeton University Press, 1985).

5. E. Majorana and L. Maiani, "A Symmetric Theory of Electrons and Positrons," in *Ettore Majorana Scientific Papers,* ed. G. F. Bassani and Council of the Italian Physical Society (Berlin: Springer, 2006), https://doi.org/10.1007/978-3-540-48095-2_10.

6. Thomas Kuhn, *The Copernican Revolution: Planetary Astronomy in the Development of Western Thought*, vol. 16 (Cambridge, MA: Harvard University Press, 1985).

7. Erasmo Recami, *The Majorana Case* (Hackensack, NJ: World Scientific, 2020).

8. Ester Palma, "La Procura: Ettore Majorana vivo in Venezuela fra il 1955 e il 1959," *Corriere Della Sera* 92 (February 2015), https://roma.corriere.it/notizie/cronaca/15_febbraio_04/procura-ettore-majorana-vivo-venezuela-il-1955-1959-d1a6aeda-ac7f-11e4-88df-4d6b5785fffa.shtml.

Chapter 5

1. Arthur Conan Doyle, *The Complete Sherlock Holmes*, (New York: Doubleday & Company, Inc, 1930), 111.

2. "Lead" in this case is pronounced like "leed" and is derived from an old term for an outcropping of gold-bearing ore.

3. Daily Digest, Office of Current Intelligence, Central Intelligence Agency, April 20, 1951, https://www.cia.gov/readingroom/docs/CIA-RDP79T01146A000100440001 -3.pdf.

4. Central Intelligence Reading Room, https://www.cia.gov/readingroom/docs/CIA -RDP80-00809A000700070611-4.pdf.

5. V. N. Gribov and B. Pontecorvo, "Neutrino Astronomy and Lepton Charge," *Physics Letters B* 28 (1969): 493.

6. B. Pontecorvo, "Electron and Muon Neutrinos," *Zhur. Eksptl'. i Teoret. Fiz.* 37 (1959).

7. A. De Rújula and S. L. Glashow, "Neutrino Helioseismology," August 11, 1992, arXiv:hep-ph/9208223.

8. Unless otherwise noted, all quotes from Sheldon Glashow come from an interview by phone, September 23, 2021.

9. S. L. Glashow, "The Future of Elementary Particle Physics," in *Quarks and Leptons*, ed. M. Lévy, J. L. Basdevant., D. Speiser, J. Weyers, R. Gastmans, and M. Jacob (Boston: Springer, 1980), https://doi.org/10.1007/978-1-4684-7197-7_15.

10. F. Reines, "Atmospheric Neutrinos as Signal and Background," in *16th International Cosmic Ray Conference, Kyoto, Japan, August 6–18, 1979, Conference Papers* (Tokyo: University of Tokyo, 1980), 60–79.

11. B. Pontecorvo, "Lepton Charges and Lepton Mixing," in *Proceedings of European Conference on Particle Physics* (Budapest: Joint Institute for Nuclear Research, 1977), 1081–1108.

12. With the possible exception of Bruno Pontecorvo.

Chapter 6

1. Franz Kafka, *The Metamorphosis and Other Stories* (New York: Barnes & Noble, 1996).

2. J. A. Formaggio, D. I. Kaiser, M. M. Murskyj, and T. E. Weiss, "Violation of the Leggett-Garg Inequality in Neutrino Oscillations," *Physical Review Letters* 117, no. 5 (2016): 050402, arXiv:1602.00041.

3. Eight MeV.

4. Michael Burghard Smy (for the Super-Kamiokande Collaboration), "The Solar Neutrino Day/Night Effect in Super-Kamiokande," *Nuclear Physics Proceedings Supplements* 138 (2005): 91–93, arXiv:hep-ex/0310064.

5. $\Delta m^2(2,1) = 7.5 \times 10^{-5} \ (eV/c^2)^2$.

6. $\Delta m^2(3,2) = 2.5 \times 10^{-3}$ $(eV/c^2)^2$.

7. C. Athanassopoulos et al., "Candidate Events in a Search for $\bar{v}_\mu \to \bar{v}_e$ Oscillations," *Physical Review Letters* 75, no. 2650 (1995); C. Athanassopoulos et al., "Evidence for $\bar{v}_\mu \to \bar{v}_e$ Oscillations from the LSND Experiment at the Los Alamos Meson Physics Facility," *Physical Review Letters* 77, no. 3082 (1996); C. Athanassopoulos et al., "Results on $\bar{v}_\mu \to \bar{v}_e$ Neutrino Oscillations from the LSND Experiment," *Physical Review Letters* 81, no. 1774 (1998); C. Athanassopoulos et al., "Results on $\bar{v}_\mu \to \bar{v}_e$ Oscillations from Pion Decay in Flight Neutrinos," *Physical Review C* 58, no. 2489 (1998); A. Aguilar et al., "Evidence for Neutrino Oscillations from the Observation of $\bar{v}_e$ Appearance in a $\bar{v}_\mu$ Beam," *Physical Review D* 64, no. 112007 (2001).

8. A. A. Aguilar-Arevalo et al. (MiniBooNE Collaboration), "Updated MiniBooNE Neutrino Oscillation Results with Increased Data and New Background Studies," *Physical Review D* 103, no. 052002 (2021).

9. $\Delta m^2(3,2) = 2.5 \times 10^{-3}$ $(eV/c^2)^2$.

Chapter 7

1. Gordon Kane, *The Particle Garden: Our Universe as Understood by Particle Physicists* (New York: Basic Books, 1996).

2. Steven Weinberg, "The Making of the Standard Model," *The European Physical Journal C-Particles and Fields* 34, no. 1 (2004): 5–13.

3. The Standard Model does not deal at all with gravity, which is important at large distances but negligible in the tiny world of elementary particles.

4. We refer to charges measured in terms of the magnitude of the electron's charge. In which case, the electron charge is defined as –1, and the W-plus has a charge +1 that is the same magnitude, but opposite in sign, as the electron charge.

5. At very high temperatures, characteristic of the early universe, quarks and gluons can be liberated to form a quark-gluon plasma. Accelerating heavy ions to very high energy, physicists have managed to create small regions of quark-gluon plasma in the laboratory

6. The spin number indicates how many units of angular momentum a particle carries. It's measured in terms of the *reduced Planck's constant* that is denoted $\hbar$. The electron and neutrinos, for example, have $\hbar/2$ units of angular momentum, while the photon has angular momentum $\hbar$. For this book, we typically omit the $\hbar$, writing 1/2, 1, or 2 in place of $\hbar/2$, $\hbar$, or $2\hbar$.

7. In the appendix, we recount some of the history of how this was discovered.

8. A description of how the mechanism works within the Standard Model is included in the appendix.

9. Oddly enough, the ball is much heavier than the players, for neutrinos interacting through the weak force.

10. Why different particles acquire radically disparate amounts of mass from the Higgs mechanism is one of the greatest mysteries of the Standard Model. If you can explain it, you will become rich and famous.

11. "What a Mathematician Learned from Cracking the Zodiac Killer's Code," Vice (n.d.). Retrieved February 20, 2022, from https://www.vice.com/en/article/k7a3gz /what-a-mathematician-learned-from-cracking-the-zodiac-killers-code.

Chapter 8

1. Planck 2018 Results—VI. Cosmological Parameters, Planck Collaboration, N. Aghanim, et al., *Astronomy and Astrophysics* 641, no. A6 (2020), doi: 10.1051/0004 –6361/201833910.

2. Guillermo F. Abellán, Zackaria Chacko, Abhish Dev, Peizhi Du, Vivian Poulin, and Yuhsin Tsai, "Improved Cosmological Constraints on the Neutrino Mass and Lifetime," arXiv:2112.13862 [hep-ph].

3. Including the particles that make up this book.

4. A. D. Sakharov, "CP Symmetry Violation, C-Asymmetry and Baryonic Asymmetry of the Universe," *Pisma Zh. Eksp. Teor. Fiz.* 5 (1967): 32.

5. Physicists discuss this in terms of charge symmetry, which they call C for short, and charge-parity symmetry, which they denote CP.

6. K. Abe and the T2K Collaboration, "Constraint on the Matter-Antimatter Symmetry-Violating Phase in Neutrino Oscillations," *Nature* 580, no. 7803 (2020): 339–344, https://doi.org/10.1038/s41586-020-2177-0.

Chapter 9

1. John N. Bahcall and Raymond Davis, "The Evolution of Neutrino Astronomy," *Publications of the Astronomical Society of the Pacific* 112.770 (2000): 429.

2. The SuperNova Early Warning System acronym was a play on the Ballistic Missile Early Warning System (BMEWS) that had been developed during the Cold War to provide alerts in the event that one of the other world superpowers were to launch a nuclear attack on the United States.

3. Frederick Reines, "Neutrinos from the Atmosphere and Beyond," *AIP Conference Proceedings* 169, no. 65 (1988), https://doi.org/10.1063/1.37194.

4. A googolplex equals $((10)^{10})^{100}$.

5. Interview by phone, June 22, 2021.

6. Interview conducted over email in June and October 2021.

7. Unless particles called tachyons exist that can travel faster than light. So far, none has been found, although neutrinos are one possible, very long shot, tachyon candidate.

8. J. G. Learned and S. Pakvasa, "Detecting v_τ Oscillations at PeV Energies," *Astroparticle Physics* 3, no. 3 (1995): 267.

9. R. Abbasi, M. Ackerman, J. Adams, J. A. Aguilar, M. Ahlers, M. Ahrens, C. Alispach, et al., "Measurement of Astrophysical Tau Neutrinos in IceCube's High-Energy Starting Events," November 6, 2020, arXiv:2011.03561.

10. Kelen Tuttle, "Ultra-High-Energy Neutrinos," *Symmetry*, November 8, 2013, https://www.symmetrymagazine.org/article/november-2013/ultra-high-energy -neutrinos.

11. O. Balducci, S. Hofmann, and A. Kassiteridis, "Cosmological Singlet Diagnostics of Neutrinophilic Dark Matter," *Physical Review D* 98, no. 023003 (2018).

12. Dailymail.com, "Researchers in Antarctica Gather Evidence of a Mirror Universe to Our Own Where Time Runs Backwards," *Daily Mail Online*, May 19, 2020, https://www.dailymail.co.uk/sciencetech/article-8338073/Researchers-Antarctica -gather-evidence-mirror-universe-time-runs-backwards.html, accessed February 26, 2022; Jon Cartwright, "We May Have Spotted a Parallel Universe Going backwards in Time," *New Scientist*, April 8, 2020, https://www.newscientist.com/article/mg245 32770-400-we-may-have-spotted-a-parallel-universe-going-backwards-in-time.

13. R. Mitalas and K. R. Sills, "On the Photon Diffusion Time Scale for the Sun," *The Astrophysical Journal* 401 (1992): 759.

14. Borexino Collaboration, "Neutrinos from the Primary Proton-Proton Fusion Process in the Sun," *Nature* 512 (2014): 383–386, https://doi.org/10.1038/nature13702.

15. M. Agostini et al. (Borexino Collaboration), "Simultaneous Precision Spectroscopy of pp, ^{7}Be, and pep Solar Neutrinos with Borexino Phase-II," *Physical Review D* 100, no. 082004 (2019).

16. W. C. Haxton, "Proposed Neutrino Monitor of Long-Term Solar Burning," *Physical Review Letters* 65 (1990): 809.

17. The IceCube Collaboration, "Detection of a Particle Shower at the Glashow Resonance with IceCube," *Nature* 591 (2021): 220–224, https://doi.org/10.1038 /s41586-021-03256-1.

18. M. Bauer and J. D. Shergold, "Observing Relic Neutrinos with an Accelerator Experiment," 2021, arXiv:2104.12784.

19. Paul Langacker, Jacques P. Leveille, and Jon Sheiman, "On the Detection of Cosmological Neutrinos by Coherent Scattering," *Physical Review D* 27, no. 1228 (1983).

20. Interview with Francis Halzen conducted by phone, June 18, 2021.

21. SN1987A was a star before it went supernova. The neutrinos that turned up in terrestrial detectors technically came from the explosion, not the star that preceded the supernova.

Chapter 10

1. C. J. Emlyn-Jones and William Preddy, eds., *Plato, Republic,* vol. 1 (Cambridge, MA: Harvard University Press, 2013).

2. G. Danby, J.-M. Gaillard, K. Goulianos, L. M. Lederman, N. Mistry, M. Schwartz, and J. Steinberger, "Observation of High-Energy Neutrino Reactions and the Existence of Two Kinds of Neutrino," *Physical Review Letters* 9, no. 36 (1962).

3. The T2K Collaboration, "Constraint on the Matter-Antimatter Symmetry-Violating Phase in Neutrino Oscillations," *Nature* 580 (2020): 339–344, https://doi.org/10.1038/s41586-020-2177-0.

4. M. A. Acero, P. Adamson, L. Aliaga, N. Anfimov, A. Antoshkin, E. Arrieta-Diaz, L. Asquith, et al., "An Improved Measurement of Neutrino Oscillation Parameters by the NOvA Experiment," August 18, 2021, arXiv:2108.08219 [hep-ex].

5. Minerva was also involved in the first transmission of a message using neutrinos as the medium of communication (see chapter 15).

6. Adrian Cho, "Costs Balloon for U.S. Particle Physics Megaproject," *Science*, September 21, 2021.

7. Jean-Pierre Delahaye, C. Ankenbrandt, Stephen Brice, Alan David Bross, Dmitri Denisov, Estia Eichten, Stephen Holmes, et al. "A Staged Muon Accelerator Facility for Neutrino and Collider Physics," February 5, 2015, arXiv preprint arXiv:1502.01647.

8. Paul Kyberd, "How a 'Muon Accelerator' Could Unravel Some of the Universe's Greatest Mysteries," The Conversation, February 11, 2020, https://theconversation.com.

9. Patrick Huber, "Neutrino Physics at a Neutrino Factory," talk presented at the Muon Collider Physics Workshop, Virginia Tech, November 30–December 2, 2020, https://indico.cern.ch/event/969815/contributions/4098178/attachments/2154917/3634401/talk.pdf.

10. Bruce J. King, "Potential Hazards from Neutrino Radiation at Muon Colliders," *Proceedings of the 1999 Particle Accelerator Conference*, vol. 1 (New York: IEEE, 1999).

11. Colin Johnson, Gigi Rolandi, and Marco Silari, "Radiological Hazard Due to Neutrinos from a Muon Collider," *Prospective Study of Muon Storage Rings at CERN* (1998): 99–102.

12. Nikolai Mokhov and Andreas Van Ginneken, "Neutrino Radiation at Muon Colliders and Storage Rings," *Journal of Nuclear Science and Technology* 37 (suppl. 1): 172–179, doi: 10.1080/00223131.2000.10874869; Bruce J. King, "Neutrino Radiation Challenges and Proposed Solutions for Many-TeV Muon Colliders," paper presented at Studies on Colliders and Collider Physics at the Highest Energies Conference, Brookhaven National Lab (BNL), Upton, NY, September 1999.

13. Jeffrey M. Berryman, Pilar Coloma, Patrick Huber, Thomas Schwetz, and Albert Zhou, "Statistical Significance of the Sterile-Neutrino Hypothesis in the Context of Reactor and Gallium Data," *Journal of High Energy Physics* 55 (2022), https://doi.org/10.1007/JHEP02(2022)055.

Chapter 11

1. Central Intelligence Agency, "Soviet Antisubmarine Warfare: Current Capabilities and Priorities," September 1, 1972.

2. J. Weber, "Method for Observation of Neutrinos and Antineutrinos," *Physical Review C* 31 (April 1, 1985): 1468.

3. The CEvNS effect confirmed in 2017 involves neutrinos scattering off individual nuclei, not a macroscopic object.

4. Virginia Trimble, "Wired by Weber," *The European Physical Journal H* 42 (2017): 261–291, https://doi.org/10.1140/epjh/e2016-70060-5.

5. The prize was ultimately awarded to Charles Townes, who relied heavily on Weber's work, along with Soviet scientists Nikolay Basov and Alexander Prokhorov.

6. JSR-85-210, submitted to OPNAV-095 in July 1985.

7. Interview by phone, September 17, 2021.

8. D. Z. Freedman, "Coherent Effects of a Weak Neutral Current," *Physical Review D* 9 (1974): 1389.

9. "Spallation" comes from the word "spall," which means a chip from something larger. The protons in a spallation source effectively chip neutrons out of atoms.

10. See the biblical passage Joshua 6:20 describing an acoustical assault on the fortress at Jericho for one legend that might rely on the phenomenon

11. Not to be confused with the CCM company that is the premier manufacturer of hockey equipment in Canada. The fact that CCM neutrino experimentalist Richard Van de Water is a Canadian and a hockey aficionado may, or may not, be a simple coincidence.

Chapter 12

1. Interview by phone, June 18, 2021.

2. Arthur Loureiro, Andrei Cuceu, Filipe B. Abdalla, Bruno Moraes, Lorne Whiteway, Michael McLeod, Sreekumar T. Balan, et al., "Upper Bound of Neutrino Masses from Combined Cosmological Observations and Particle Physics Experiments," *Physical Review Letters* 123, no. 8 (2019): 081301.

3. Interview by phone, June 18, 2021.

4. S. Mertens, "Direct Neutrino Mass Experiments," *Journal of Physics: Conference Series* 718, no. 2 (2016): 022013.

5. Interview by phone, May 4, 2021.

Chapter 13

1. Alan Chodos, Avi I. Hauser, and V. Alan Kostelecký, "The Neutrino as a Tachyon," *Physics Letters* 150B (1985): 431.

2. In the case of negative mass squared, the mass itself is a so-called imaginary number involving the square root of negative one—but this does not make it unacceptable physically.

3. Joel Achenbach, "Faster-than-Light Neutrino Poses the Ultimate Cosmic Brain Teaser for Physicists," *Washington Post*, November 14, 2011, https://www.washingtonpost.com/national/health-science/faster-than-light-neutrino-poses-the-ultimate-cosmic-brain-teaser-for-physicists/2011/11/09/gIQAsw9sKN_story.html.

4. A. G. Cohen and S. L. Glashow, "Pair Creation Constrains Superluminal Neutrino Propagation," *Physical Review Letters* 107, no. 18 (2011): 181803.

5. Gerald Holton, "Einstein, Michelson, and the 'Crucial' Experiment," *Thematic Origins of Scientific Thought*, 3rd ed. (Cambridge, MA: Harvard University Press, 1995 [1973]), 279–370.

6. V. Alan Kostelecký and Matthew Mewes, "Lorentz and CPT Violation in Neutrinos," *Physical Review D* 69 (January 30, 2004): 016005.

7. Jorge S. Díaz, "Neutrinos as Probes of Lorentz Invariance," *Advances in High Energy Physics* 2014 (2014).

8. P. deNiverville, M. Pospelov, and A. Ritz, "Observing a Light Dark Matter Beam with Neutrino Experiments," *Physical Review. D* 84 (2011): 075020.

Chapter 14

1. Interview by phone, September 20, 2021.

2. J. W. C. Mills, "There's No Replacement for Displacement," Petrolicious, September 26, 2013, https://petrolicious.com/articles/there-s-no-replacement-for-displacement, accessed February 28, 2022.

3. M. Bowen and P. Huber, "Reactor Neutrino Applications and Coherent Elastic Neutrino Nucleus Scattering," *Physical Review D* 102 (2020): 053008.

4. A. Bernstein, N. Bowden, B. L. Goldblum, P. Huber, I. Jovanovic, and J. Mattingly, "Colloquium: Neutrino Detectors as Tools for Nuclear Security," *Reviews of Modern Physics* 92 (2020): 011003.

5. M. Lindner, T. Ohlsson, R. Tomàs, and W. Winter, "Tomography of the Earth's Core Using Supernova Neutrinos," *Astroparticle Physics* 19 (2003): 755.

6. A. De Rújula, S. L. Glashow, R. R. Wilson, and G. Charpak, "Neutrino Exploration of the Earth," *Physics Reports* 99 (1983): 341.

7. A. Donini, S. Palomares-Ruiz, and J. Salvado, "Neutrino Tomography of Earth," *Nature Physics* 15 (2019): 37–40, https://doi.org/10.1038/s41567-018-0319-1.

8. N. P. Pitjev and E. V. Pitjeva, "Constraints on Dark Matter in the Solar System," *Astronomy Letters* 39 (2013): 141–149, https://doi.org/10.1134/S1063773713020060.

9. T. Ohlsson and W. Winter, "Could One Find Petroleum Using Neutrino Oscillations in Matter?," *Europhysics Letters* 60 (2002): 34–39.

10. Walter Winter, "Probing the Absolute Density of the Earth's Core Using a Vertical Neutrino Beam," *Physical Review D* 72 (August 2005): 037302.

11. J. Perry, "On the Age of the Earth," *Nature* 51 (1895): 224–227, 582–585.

12. S. T. Dye, "Geoneutrinos and the Radioactive Power of the Earth," *Reviews of Geophysics* 50 (2012): RG3007, https://doi.org/10.1029/2012RG000400.

13. J. Marvin Herndon, "Nuclear Georeactor Generation of Earth's Geomagnetic Field," July 19, 2007, arXiv:0707.2850.

14. F. Reines, R. L. Schuch, C. L. Cowan, F. B. Harrison, E. C. Anderson, and F. N. Hayes, "Determination of Total Body Radioactivity Using Liquid Scintillation Detectors," *Nature* 172 (1953): 521–523, https://doi.org/10.1038/172521a0.

15. In 1981, the facility name was changed to Los Alamos National Laboratory.

16. B. Al Ali, M. Garel, P. Cuny, J. C. Miquel, T. Toubal, A. Robert, and C. Tamburini, "Luminous Bacteria in the Deep-Sea Waters near the ANTARES Underwater Neutrino Telescope (Mediterranean Sea)," *Chemistry and Ecology* 26, no. 1 (2010): 57–72.

17. M. André, A. Caballé, M. Van der Schaar, A. Solsona, L. Houégnigan, S. Zaugg, A. M. Sánchez, et al., "Sperm Whale Long-Range Echolocation Sounds Revealed by ANTARES, a Deep-Sea Neutrino Telescope," *Scientific Reports* 7, no. 1 (2017): 1–12.

Chapter 15

1. Defense Intelligence Agency, Soviet and Czechoslovakian Parapsychology Research report, September 1, 1975, https://www.cia.gov/readingroom/document /cia-rdp96-00787r000500420001-2.

2. The Soviets were not alone in their pursuit of dubious paranormal research. For more on the U.S. programs, see Jon Ronson, *The Men Who Stare at Goats* (New York: Simon and Schuster, 2004), and David Kaiser, *How the Hippies Saved Physics: Science, Counterculture, and the Quantum Revival* (New York: W. W. Norton, 2011).

3. A. W. Sáenz, H. Überall, F. J. Kelly, D. W. Padgett, and N. Seeman, "Telecommunication with Neutrino Beams," *Science* 198, no. 4314 (October 21, 1977).

4. Interview by phone, February 3, 2021.

5. D. D. Stancil, "Demonstration of Communication Using Neutrinos," *Modern Physics Letters A* 27 (2012): 1250077.

6. C. Coupé, Y. Oh, D. Dediu, and F. Pellegrino, "Different Languages, Similar Encoding Efficiency: Comparable Information Rates across the Human Communicative Niche," *Science Advances* 5, no. 9(2019): eaaw2594.

7. G. Laughlin, A. Aguirre, and J. Grundfest, "Information Transmission between Financial Markets in Chicago and New York," *Financial Review* 49 (2014): 283–312.

8. Paul Davies and Richard Carrigan, "The Eerie Silence: Renewing Our Search for Alien Intelligence," *Physics Today* 63, no. 10 (2010): 55.

9. M. Abramowicz, M. Bejger, É. Gourgoulhon, and O. Straub, "A Galactic Centre Gravitational-Wave Messenger," *Scientific Reports* 10, no. 7054 (2020), https://doi .org/10.1038/s41598-020-63206-1.

10. John G. Learned, Sandip Pakvasa, and A. Zee, "Galactic Neutrino Communication, *Physics Letters B* 671, no. 1 (2009): 15–19.

11. Interview by phone, September 17, 2021.

12. M. Hippke, "Benchmarking Information Carriers," *Acta Astronautica* 151 (2018): 53–62.

13. John G. Learned, R.-P. Kudritzki, Sandip Pakvasa, and A. Zee, "The Cepheid Galactic Internet," *Contemporary Physics* 53, no. 2 (2012): 113–118.

14. Brian C. Lacki, "SETI at Planck Energy: When Particle Physicists Become Cosmic Engineers," March 6, 2015, https://arxiv.org/pdf/1503.01509.pdf.

15. Freeman J. Dyson, "Search for Artificial Stellar Sources of Infrared Radiation," *Science* 131, no. 3414 (1960): 1667–1668.

Chapter 16

1. F. Reines and C. L. Cowan, "The Neutrino," *Nature*, September 1, 1956.

Further Reading

Asimov, Isaac. *The Neutrino: Ghost Particle of the Atom*. Garden City, NY: Doubleday, 1966. The first popular introduction to the neutrino by one of the greatest writers of the twentieth century.

Bible. Book of Joshua, chapter 6. Available in many editions and many languages.

Bowen, Mark. *The Telescope in the Ice: Inventing a New Astronomy at the South Pole*. New York: St. Martin's Press, 2017. An eyewitness account of how IceCube came to be.

Carroll, Sean. *The Particle at the End of the Universe: How the Hunt for the Higgs Boson Leads Us to the Edge of a New World*. New York: Dutton, 2012. The Standard Model and why there was excitement over the Higgs boson discovery.

Close, Frank. *Elusive: How Peter Higgs Solved the Mystery of Mass*. New York: Basic Books, 2022. A joint biography of British physicist Peter Higgs and the particle named after him.

Close, Frank. *Half Life: The Divided Life of Bruno Pontecorvo, Physicist or Spy*. London: Oneworld Publications, 2015. A detailed account of the life and science of Bruno Pontecorvo.

Davies, Paul. *The Eerie Silence: Renewing Our Search for Alien Intelligence*. Boston: Mariner Books, 2011.

Farmelo, Graham. *The Strangest Man: The Hidden Life of Paul Dirac, Mystic of the Atom*. New York: Basic Books, 2009. Dirac was one of the foremost quantum pioneers with uncanny creative mathematical abilities.

Kafka, Franz. *The Metamorphosis*. Available in many editions. Neutrinos are not mentioned.

Kaiser, David. *How the Hippies Saved Physics: Science, Counterculture and the Quantum Revival*. New York: W. W. Norton, 2011. All about quantum weirdness and beyond.

Magueijo, Joao. *A Brilliant Darkness: The Extraordinary Life and Mysterious Disappearance of Ettore Majorana, the Troubled Genius of the Nuclear Age*. New York: Basic Books, 2009. The ever-fascinating story of Ettore Majorana.

Pais, Abraham. *Inward Bound: Of Matter and Forces in the Physical World*. Oxford: Oxford University Press, 1986. A history of particle physics from the late nineteenth century to the mid-1980s. Warning: Contains equations.

Peierls, R. *Wolfgang Ernst Pauli*. London: Biographical Memoirs of the Royal Society, 1960. https://royalsocietypublishing.org/doi/pdf/10.1098/rsbm.1960.0014. A remembrance of Wolfgang Pauli by a physicist who knew him well, with interesting anecdotes at the end.

Recami, Erasmo. *The Majorana Case: Letters, Documents, Testimonies*. Singapore: World Scientific, 2020. More about Majorana.

Segrè, Gino, and Bettina Hoerlin. *The Pope of Physics: Enrico Fermi and the Birth of the Atomic Age*. New York: Henry Holt and Co., 2016. Fermi was outstanding as both a theorist and an experimentalist, possessed of legendary physical intuition.

Index